Process Development Division 2018

Core Programming Area at the 2018 AIChE Annual Meeting

Pittsburgh, Pennsylvania, USA
28 October - 2 November 2018

ISBN: 978-1-5108-7622-4

Process Development Division 2018

Core Programming Area at the 2018 AIChE Annual Meeting

Pittsburgh, Pennsylvania, USA
28 October - 2 November 2018

AIChE
The Global Home of Chemical Engineers

Printed from e-media with permission by:

Curran Associates, Inc.
57 Morehouse Lane
Red Hook, NY 12571

CURRAN ASSOCIATES INC.
proceedings.com

Some format issues inherent in the e-media version may also appear in this print version.

Printed by Curran Associates, Inc. (2019)

AIChE
120 Wall Street, FL 23
New York, NY 10005-4020

Phone: (800) 242-4363
Fax: (203) 775-5177

www.aiche.org

Additional copies of this publication are available from:

Curran Associates, Inc.
57 Morehouse Lane
Red Hook, NY 12571 USA
Phone: 845-758-0400
Fax: 845-758-2633
Email: curran@proceedings.com
Web: www.proceedings.com

TABLE OF CONTENTS

Some abstracts in this book refer to images that were not made available by the publisher

535550 Development of Nanoparticle Alignment Regimes in Drying Cellulose Nanocrystal Droplet Suspensions for Additive Manufacturing

Abstract:

Nanoparticle suspensions have potential for a number of drop-on-demand applications, where they are deposited in droplet form and subsequently dried via evaporation of the suspending agent. The droplet drying process often results in deposition of ring patterns, typically referred to as the coffee ring effect. When drying picoliter-sized, aqueous, cellulose nanocrystal droplet suspensions, which are on the size scale of traditional drop-on-demand technologies, we observe alignment of these rod-like nanoparticles in multiple orientations. A radial alignment is formed near the center of the droplet, and tangential orientation is observed in the particles along the circumference of the droplet. Tuning particle orientation enables unique opportunities for control of nanocomposite structure required for next generation optical and electronic materials that can be generated using drop-on-demand or voxel based technologies. To successfully control the nanoparticle orientation during the drying process, a fundamental understanding of the alignment mechanism is necessary.

We have developed a 3-dimensional, continuum-based, finite element model implementing a deforming mesh to understand the impact of shear induced alignment driven by evaporative flow. Simulations of a single-phase, sessile water droplet suggest a relationship between velocity profile, shear stress, and particle orientation. We have explored the coupled effects of Marangoni flows, air convection, heat transport, and concentration dependent viscosity models on evaporation of single-phase picoliter-sized droplets leading to new insights into particle orientation in the evaporation of colloidal droplets that result from a complex interplay of initial concentration, substrate surface functionality, and particle-particle interactions. Velocity and shear stress profiles indicate the development of radial flows enacting shear stresses on the fluid driving the alignment of rod-like nanoparticles in the flow direction which varies with location in the confined flow field.

522259 Characterization of Poly(ether imide) Towards the Development of a Fused Filament Fabrication (FFF) Process Model

Abstract:

Fused filament fabrication (FFF) is the most common method of AM that relies on thermoplastic extrusion. As the AM field is transitioning from prototype visualization to functional prototypes and end product development, one of the greatest hurdles is the inherent z-axis (interlayer) weakness of the processed parts resulting from weak adhesion between deposited layers, producing anisotropic mechanical properties. This poor adhesion stems from insufficient polymer inter-diffusion between the layers, referred to as â€œself-healing.â€

To better understand the evolution of part strength during the FFF process, we are developing a model to describe the self-healing phenomenon that couples the complex thermal and geometric profiles with diffusion at the layer interface. This model is strongly dependent on the transient behavior of the polymer governed by the processing and environmental conditions present during the printing process. Additionally, we are examining the effect of adding a continuous composite fiber with anisotropic thermal conductivity. Our preliminary work has developed a realistic three-dimensional model of a printed part which will accurately represent the interface between individual layers. Previous models have utilized idealistic models based on circles and ovals of equal size. Our results prove the individual layers are flattened ellipsoids that vary in diameter depending on their height from the heated bed. We have also analyzed the complete re-entanglement time of the neat polymer to quantify the extent of self-healing, combined with characterizing the thermal conductivity of the composite filaments. The corresponding mechanical strength is driven by interlayer adhesion and characterized as a function of processing conditions. Our preliminary rheological findings of a poly(ether imide) amorphous thermoplastic indicate that times for complete re-entanglement range from 47 seconds to more than 22 minutes for isothermal temperatures of 375 °C and 345 °C, respectively. Since the cooling process is non-isothermal, and the layers remain above their glass transition temperatures for less than 3 seconds, this would limit the amount of chain re-entanglement at the layer interface and corresponding bond strength between layers of a printed part, and suggests that a modified printing process is required to create stronger parts with this material. The inclusion of the composite fiber does increase the time at which the temperature of the polymer is elevated and should help increase the amount of self-healing that occurs during the printing process.

537447 Additive Manufacturing of Core-Shell Microparticles Containing Thermosetting Resins

Abstract:

Cold spray deposition is a common additive manufacturing method for fabricating metal coatings that involves the spraying metal particles at high-speeds. However, this scalable, additive manufacturing process has not yet generated solid polymer coatings. Our study enables the cold spraying of solid polymer materials at room temperature. Here, we developed novel, core-shell thermoplastic microparticles containing a high concentration of thermosetting resins. Suspension polymerization was employed to produce the microparticles due to its many advantages, namely low conversion cost, mature scale-up processing, and a versatile tunability in terms of particle size and morphology. The thermal properties, size distribution, morphology, and composition of the particles were explored and fine-tuned in an effort to optimize the deposition of the particles at room temperature using cold spray technology. The outcome of this study is a novel form of sprayable thermosetting powders that deposit to form a coating with a high performance, easy operation processing, significant environmental benefits, and a commercial feasibility.

532737 Supersonic-Impaction Printing of Flame-Made Doped-Perovskite Nanoparticles

Abstract:

Perovskite nanoparticles find a range of applications in emerging electronics such as transparent conductors, solar cells, light-emitting diodes and sensors such as gas sensing. Such electroactive oxides typically have melting temperatures over 1000°C. Subsequently, their deposition, calcination and sintering also require extremely high temperatures. As a result, substrate compatibility with high temperature treatments is needed to facilitate usability and implementation of such materials. This is therefore a bottleneck for their application on substrates that are susceptible to high temperatures and oxidizing environments such as plastics or metals. More recently, the advent of additive manufacturing has shed new light on the deposition and growth of oxide materials. Among them, perovskite materials synthesized via sol-gel methods followed by inkjet and aerosol printing has shown promise for being able to be deposited on low temperature substrates. In this presentation, we will be showcasing a new concept and design for a deposition chamber intended for the supersonic impaction of aerosolized nanomaterials. Further, we will also discuss the implementation of a flame spray reactor as the source of synthesizing pure Barium Stannate ($BaSnO_3$) nanoparticles that are impacted using supersonic deposition to create electrically conducting perovskite thin films.

The principle of supersonic deposition is primarily based on acceleration of an aerosol-containing gas stream through a converging-diverging (de Laval) nozzle. This de Laval nozzle creates a high pressure drop across itself when downstream is pumped to moderate vacuum and upstream is maintained at a higher pressure. The pressure gradient across the nozzle accelerates the flowing aerosol to supersonic speeds and subsequently impact them onto the substrate. The kinetic energy of the particles is converted into heat upon impaction and thus the process can create a homogeneous thin film of sintered materials. Through computational fluid dynamics simulations, we found a number of parameters including the effect of gas velocities, nanoparticle density, size, trajectories and other physical properties to affect the efficiency of impaction. Unlike other precursors for competing additive manufacturing techniques such as inkjet, aerosol and electro-hydrodynamic jet printing, supersonic impaction does not require the use of any surfactant stabilized nanoparticle inks.

For this reason, flame spray pyrolysis was chosen as a scalable route for the synthesis of pure aerosolized ceramic nanoparticles without the use of any surfactants. More recently, flames are also being implemented for the synthesis of complex oxides such as tertiaries and perovskites. We found that the critical parameters that affect the flame temperature are the precursor and dispersion oxygen flow rates and enthalpy of the precursor material. A higher feed rate and precursors with high enthalpies ensure higher temperatures of the flame. Doped $BaSnO_3$ is an electroactive perovskite material and has a large indirect intrinsic band-gap of 3.2 eV. We will discuss our developments pertaining to the incorporation of electrical properties in the thin films of $BaSnO_3$ *via* vacuum annealing to induce oxygen vacancies and incorporation of external dopants that can substitute the Ba^{2+} site (n-type with La) or Sn^{4+} sites (p-type with K, Cs).

541761 Simulating Powder Handling Processes in Additive Manufacturing Using the Discrete Element Method

Abstract:

Additive Manufacturing is a rapidly advancing manufacturing area that enables users to produce complex parts in a short amount of time. One common approach to AM in industry involves the deposition and selective sintering/melting of successive layers of metallic powders. Through repeated build-up of powder layers, complex metal parts can be produced.

There are a wide range of metal powders available for AM production and each has its own characteristics. During printing a powder must be stored, transported, and processed within the printing equipment to produce the final part. Each of these stages need to be performed efficiently by the printer to ensure an optimum printing process.

The specific composition and behavioral traits of the powder used in AM is highly influential on printer set-up and resulting part manufacturing quality. Introducing a new powder into an AM process means that the set-up of the printing device needs to be configured to handle this new material effectively. Because of the variability of powder properties, flowability and susceptibility to environmental conditions, it is not straight-forward to make decisions as to how to reach the optimal configuration. At present, a common method to adjust printer set-up to account for the influence of a specific powder requires trial and error testing, or a reliance on know-how from previous work with the same printer and powder combinations. This takes time and extends the overall part production cycle.

Discrete Element Modelling (DEM) is a simulation method for recreating bulk and granular material behaviors in a virtual environment. It is a method used in other powder-focused industries, such as pharmaceutical, to provide engineering insight into powder handling and processing operations. By using DEM simulation, engineers can test design and process concepts and extract meaningful information from virtual versions of their material and equipment to help in the decision-making process.

The same opportunities exist in AM, and such a simulation method offers engineers an alternative to physical trial and error testing when trying to understand how key stages in part production will perform. In this presentation we will discuss the application of DEM, using EDEM software, to the additive manufacturing print process. Examples will be given to demonstrate how DEM can be used to help engineers understand the performance of powder delivery systems, as well as gaining insight into metrics such as powder porosity variation during the powder-layering process.

526233 Effects of Solvent Selection on Efficient Furfural Production

Abstract:

Conversion of 5-carbon (C_5) sugars in biomass to furfural is an early step in the value chain to producing renewable value-added chemicals such as furfural alcohol, methyl tetrahydrofuran, tetrahydrofuran, 1,4 butanediol, and other four and five-carbon chemical intermediates. Molar yields of furfural from C_5 sugars are limited by competitive reactions to form humins. The commercially practiced technology for furfural production involves the simultaneous conversion of C_5 sugars in solid biomass and steam stripping of the resulting furfural. The process has a molar yield between 45-55 mol%.

Recently, researchers have examined more efficient methods of converting aqueous C_5 sugars, generally available as byproduct streams from pulp and paper manufacturing, into furfural in a two-liquid phase process. In this process, a water-immiscible organic solvent and acid catalyst are added to the aqueous C_5 stream. The organic solvent is used to selectively partition furfural away from the acid catalyst in the aqueous phase, thereby reducing the formation of humins and increasing furfural selectivity.

In this talk we will discuss the overall process considerations used to select an optimized solvent and the effect of the solvent on the energy load on distillation column used to separate furfural from the organic solvent. We will discuss the how varying the 4-tert-butyl phenol content in solvent and the organic to aqueous phase ratio affected the xylose conversion rate. We observed a 3x increase in reaction rate over the parameter space explored. The likely reasons and the implications of this finding will be discussed.

526657 Synthesis & Characterization of Molecularly Hybrid Bisphenols Derived from Lignin & Cashew Nutshell Liquid: Resin and Polymer Properties

Abstract:

Vinyl Ester (VE) resins can be used to create polymers with favorable properties for high performance applications. Standard industrial VE resins are comprised of monomers with a bisphenolic structure, which result in polymers with high strengths, moduli, thermal stabilities, and chemical resistances. Most of the commercially available VE resins are derived from bisphenol A (BPA). Unfortunately, BPA is considered unsustainable because it is synthesized from petrochemical feedstocks. It has also been scrutinized for its toxicity to the human endocrine system. Bio-based alternatives are being investigated to mitigate the negative aspects of BPA.

Lignin, a bio-based resource, is a favorable feedstock for the synthesis of VE resins due to its extensive aromaticity. Lignin is the second most abundant natural resource and a major byproduct of the pulp and paper industry. Vanillin can be derived from lignin and further reduced to vanillyl alcohol, a viable platform chemical for the production of bisphenols. Cashew nutshell liquid (CNSL) is an aromatic, renewable resource produced from the inedible shells of cashew plants. CNSL contains many phenolic lipids, the primary of which is cardanol. Cardanolâ€™s C15 alkyl chain, which can be naturally saturated or unsaturated, may impart water and thermal resistant properties to the resultant polymer network.

This study shows the synthesis and properties of lignin-derived vanillyl alcohol coupled with CNSL-derived cardanol (VAC). The VAC is an asymmetric bisphenol that is then functionalized to produce a dimethacrylate (VACDM). All dimethacrylates used in this work were blended with a reactive diluent, analyzed for their rheological properties, and cured. The resulting polymers were characterized via FT-IR, TGA, DMA, and fracture toughness to quantify the thermomechanical, thermogravimetric, and mechanical properties. The bio-based dimethacrylates were compared to BPA-derived industrial standard Vinyl Ester 828 (VE828) and the structurally similar bisphenol A dimethacrylate (BPADM).

535374 Synthesis, Design and Thermodynamic Analysis of Hybrid Processes Gasifying Biomass and Smelting Iron

Abstract:

Emerging research has revealed tremendous benefits in hybridizing the processes of iron ore reduction and biomass gasification. The iron ore acts as both an oxidant and a catalyst for the biomass gasification, while the biomass in turn replaces coking coal as the reducing agent for iron ore. The result is a carbon neutral iron ore reduction process co-producing synthesis gas.

This research presents detailed thermodynamic analysis of this system of chemical reactions to develop process targets and offer a starting point for design and optimization, as well as a basis for economic evaluation. Also presented is a sensitivity analysis to determine the effect of silica content on process performance.

It has been found that there is an inverse linear proportionality between iron reduction capacity and synthesis gas production and that the best economic performance under present market conditions is achieved by maximizing iron smelting and producing little or no synthesis gas.

However, the capacity for co-production allows for the operational flexibility to operate at maximum capacity during periods of reduced iron throughput, by transitioning to the production of synthesis gas. Once an iron deposit is depleted, a smelter capable of biomass gasification need not be scrapped; it can instead transition to full-time fuel production. In this way, iron/synthesis gas co-production can be thought of as a stepping stone to long-term renewable fuel production.

534805 F D M E (2,5-Furandicarboxylate, DiMethyl Ester) Process Development: Scale-up through Pilot

Abstract:

DuPont Industrial Biosciences and the Archer Daniels Midland (ADM) company are in a strategic partnership to develop and commercialize a new process to produce renewable FDME from corn-derived fructose. This newly commercial monomer is the dimethyl ester of 2,5-furan dicarboxylic acid (FDCA), which along with existing renewable diols such as BioPDO (1,3-propanediol) and ethylene glycol, can be polymerized to make a new family of bio-based furanic polymers (DuPont Everact^{TM}) including polytrimethylene furandicarboxylate (PTF) and polyethylene furandicarboxylate (PEF). These polyesters are of interest because of their enhanced gas barrier properties relative to incumbent polymers used in the beverage, food, and industrial packaging markets such as poly(ethylene terephthalate) (PET), and they provide opportunity for improved shelf life and decreased overall material use in packaging applications where O_2 and CO_2 permeation is critical.

This presentation will provide an overview of the process as well as the scale-up methodology being employed to develop an economically-viable FDME manufacturing process. The presentation will include a summary of best practices incorporated into the design phase such as internal third-party technical reviews of conceptual process designs and development of a dynamic process model to develop control strategies, tune process controllers, and train operators. Finally, weâ€™ll summarize the structured risk assessment and mitigation methodology employed to identify and address key technical risks to commercialization.

531142 Reuse and Valorization of Used Cooking Oils By Transformation into Epoxidized Oils

Abstract:

Currently, sustainability and circular economy are fundamental elements that are increasingly incorporated as design principles of novel chemical processes. In this direction, the use of waste residues as raw materials for the manufacture of chemical products is a clear example of these ecofriendly principles. Among different biobased residues suitable for chemical exploitation, used cooking oils (UCOs) are of importance as they represent a major problem for modern large cities. In several urban centers worldwide, UCOs are incorrectly disposed through de sewage systems, generating a cascade of problems (sewage clocking, floodings, vectors generation, bad odors, etc.) with huge environmental, social and economic impacts. Alternatively, UCOs can be used as second generation raw materials for the manufacture of different derivatives (e. g. fuels and chemicals), with the advantage of being renewable, low-cost, high-volume, readily-available, and suitable to exploit as they already delivered their nutritional value.

After pretreatment and purification processes, the UCOs can be considered as an oleochemical feedstock. It can be transformed into a large variety of byproducts such as resins, surfactants, monomers, additive for polymers, lubricants, epoxidized oils, etc. Generally, a major limitation identified when dealing with UCOs is their chemical heterogeneity due to large variety of impurities generated during the cooking process (such as aldehydes, ketones, peroxides, cylindrical and polymeric compounds). However, even after the cooking process, the UCOs maintain around 90% of its original triglycerides content, making them useful in the oleochemical industry.

In this direction, this work developed a process for the reuse and upgrade of UCOs into epoxidized oils. Samples of UCOs collected from larger generators (fast food restaurants) in the city of Bogotá were used as raw material for the process. Initially, the heterogeneity of UCOs was assessed by collecting and characterizing UCOs from different fried chicken and hamburgers restaurants along the city. The recollected samples were characterized by evaluation of acid value, content of volatile matter, density, peroxide value, color, and content of total polar compounds. This allowed to stablish the variability of the properties, and the required pretreatment process to use the UCOs as oleochemical feedstock. The range of iodine values and saponification index indicated that the collected UCOs can be used as raw materials for epoxidation. After performing filtration, degumming and bleaching processes, the UCOs were adjusted for further epoxidation.

Finally, the epoxidation of the pretreated UCOs was studied by using a traditional *in-situ* peroxyacetic acid method, with a homogeneous acid catalyst. The impact of main process variables of the epoxidation process such as temperature (50, 65, 80ºC), hydrogen peroxide excess respect acetic acid (15, 30, 45%p/p), ratio of acetic acid to UCO (1.0, 3.5, 6.0 %p/p) were assessed by a three-level factorial design, using 2%molar relation of catalyst concentration. The reaction was analyzed by measuring changes in iodine value and oxirane value. The obtained epoxidized oil exhibited oxirane oxygen content of around 3.2% with conversions above 60%. The observed epoxidation efficiency and the product specifications were similar to those obtained with virgin vegetable oils, turning the developed process into a feasible alternative for UCOs upgrading and valorization.

533694 3D Printed Complex Dosage Forms Embedded with Engineered BCS Class II Drug Particles

Abstract:

Recent work suggests complex structured dosage forms can be manufactured where drug particles can be integrated via 3D printing technology. However, incorporating engineered drug particles into 3D printed structures so that various intermediate and final product properties can be enhanced, for example, good mechanical properties and drug content uniformity of feed materials, has yet to be explored. It is hypothesized that engineered particles enable uniform distribution of drug into molten polymeric matrix via twin screw extruder (TSE). This distinct feature of engineered particles leads to obtaining enhanced dissolution profile from 3D structured dosage forms. The printable filaments containing varying drug loading (DL) %, and different extent of particle engineering as well as filament their mechanical properties were considered in this investigation. Griseofulvin (GF) and Hydroxypropyl cellulose (HPC) SL grade was used as the model BCS Class II drug and the polymeric matrix, respectively. The feed materials having four different DL% (5, 10, 15, and 25 wt.% GF) incorporating engineered and non-engineered particles were extruded by HME and 3D structured tablets from the feed materials were successfully created via FDM 3D printer. No drug degradation was observed in TGA results for any cases. All the drug particles kept their crystalline structure from HME extrusion (140 °C) as well as FDM printing (170 °C). The feed materials containing engineered particles exhibited higher tensile strength and better modulus of elasticity, regardless of the drug loading. Moreover, 3D printed tablets containing engineered particles showed enhanced drug release for all DL%. Overall, the 3D printed tablets embedded with engineered drug particles resulted in tablets without any degradation/recrystallization issues and demonstrated enhanced dissolution profiles for four different dosage amounts.

529744 3D Printing for Rapid Prototyping of Innovative Process Equipment for Pharmaceutical Crystallization

Abstract:

The development of pharmaceuticals involves various stages with preclinical and clinical trials. A short time-to-market is essential to maximize the patent protection and to avoid any loss in potential revenues.[1] Furthermore, most pharmaceuticals in development do not reach commercial manufacturing. However, for those pharmaceuticals that do reach market, efficient and novel processing equipment can provide a competitive edge. Hence, rapid prototyping of processing equipment during development is important to minimize time-to-market, while maximizing early learning on processing. 3D printing is an attractive technology to fabricate customized and novel processing equipment almost instantaneously. However, despite this potential, the use of 3D printing during early pharmaceutical process development is still in its infancy.

3D printing has found applications for the fabrication of micro-reactors and to some extent for mesoscale reactors as well.[2] In contrast to reactors, the use of 3D printing to fabricate innovative designs of separators has received limited attention. For example, for crystallization, which is a key unit operation in many pharmaceutical processes, no studies on the use of 3D printing for rapid prototyping have been reported to the best of our knowledge. Pharmaceutical crystallization is complicated by the need to optimize intrinsic crystal quality attributes such as size distribution, which affects the performance of downstream processes and possibly the efficacy of the final product. Optimization of crystallization requires control over various mixing processes. 3D printing can be a flexible and rapid method to fabricate innovative structures of crystallizers for optimized mixing at all relevant scales during pharmaceutical development.

The objective of this work is to develop and characterize 3D printed prototypes of innovative crystallization equipment with different structures for mixing. Two cases will be investigated. A common feature of both cases is that the mixing mechanism does not require moving parts, which makes fabrication with 3D printing attractive.

The first case involves a 3D-printed airlift crystallizer for protein crystallization. Protein crystallization in an airlift crystallizer has not been reported yet, but is attractive as protein crystals are generally more fragile in comparison to crystals of small-molecule active ingredients.[3] Therefore, mild and homogeneous mixing conditions are required, which can be provided by airlift columns due to the low maximum shear forces.[4] An airlift crystallizer was designed and fabricated with a desktop stereolithography (SLA) printer. The hydrodynamics of the column have been characterized for lysozyme as model protein. A sufficiently high upward velocity in the column demonstrates practical feasibility for protein crystallization in an airlift crystallizer. Next, the crystallization performance of the airlift crystallizer is characterized in this work and compared to a conventional stirred tank crystallizer.

The second case involves antisolvent and reactive crystallization in continuous flow. Rapid mixing is achieved by using static mixers. The effectiveness of kenics-type of static mixers has previously been demonstrated for anti-solvent crystallization of ketoconazole [5]. In this work, Y-mixers and komax-type mixers have been designed and fabricated using a desktop stereolithography (SLA) 3D printer. Preliminary studies on reactive crystallization of barium sulfate have been conducted, which demonstrate practical feasible and reduced agglomeration when using the komax mixer in comparison to when using a Y-mixer. The stability of many stereolithography resins used for 3D-printed structures is good for aqueous-based system. However, pharmaceutical crystallization often involves organic compounds. Therefore, this case particularly focuses on the use of various 3D printing techniques to address stability issues for organic solvents. In the long term, this research aims to develop a platform around 3D printing, which allows users to select, design and fabricate innovative crystallization equipment rapidly to achieve shorter time-to-market and intensified pharmaceutical crystallization.

Acknowledgement: The work described in this abstract was supported by a grant from the Research Grants Council of the Hong Kong Special Administrative Region, China (Project No. 16242916).

References

[1] G. Reklaitis, J. Khinast, and F. J. Muzzio, "Pharmaceutical engineering science—New approaches to pharmaceutical development and manufacturing," Chem. Eng. Sci., vol. 65, no. 21, pp. iv–vii, Nov. 2010.

[2] A. K. Au, W. Huynh, L. F. Horowitz, and A. Folch, "3D-Printed Microfluidics," Angew. Chemie - Int. Ed., vol. 55, no. 12, pp. 3862–3881, 2016.

[3] S. Tait, E. T. White, and J. D. Litster, "Mechanical Characterization of Protein Crystals", Part. Part. Syst. Charact., vol. 25, no. 3, pp. 266-276, Sep 2008.

[4] A. Soare, R. Lakerveld, J. Van Royen, G. Zocchi, A. I. Stankiewicz, and H. J. M. Kramer, "Minimization of attrition and breakage in an airlift crystallizer," Ind. Eng. Chem. Res., vol. 51, no. 33, pp. 10895–10909, Aug 2012.

[5] A. J. Alvarez and A. S. Myerson, "Continuous plug flow crystallization of pharmaceutical compounds," Cryst. Growth Des., vol. 10, no. 5, pp. 2219–2228, May 2010.

529577 Direct Write of UV Curable Polymer Bonded Magnets

Abstract:

Today, almost all commercial magnet producers are using the conventional sintering-based procedure to manufacture magnets. Polymer-bonded hard magnets have recently gained popularity as a novel magnet manufacturing technique due to its design freedom regarding shapes and magnetizing structures. Comparing to the conventional sintering process, the maximum energy product $(BH)_{max}$ of the polymer-bonded magnets is lower. However, the fabrication of polymer-bonded magnets does not require high temperature treatment and thus can be more efficient. Furthermore, no current mass production technologies can produce magnets with complex geometries.

This work utilizes extrusion-based direct write (DW), combined with UV curing, to fabricate polymer-bonded magnets at room temperature. DW is an additive manufacturing technique enabling the deposition of electronic components and functional or structural patterns using different types of materials, directly without utilizing masks or subsequent etching processes. Compared to the molding process, DW provides intricate details and smooth surface finish. In this work, Nd-Fe-B powder are directly mixed with photo-curable binders to form a printable DW ink. Experiments were carried out to understand the relationship between magnetic powder solid loading, processing, and mechanical strength. Magnetic performance of printed parts was compared to other polymer-bonded and sintered magnets. The study opens a new pathway for preparing polymer-bonded magnets at room temperature for various magnetic applications.

523345 Before You Click "Print": Regulatory Considerations for 3D Printed Oral Drug Products

Abstract:

Additive manufacturing/3D printing of oral drug products and medical devices has become more attractive to shift the manufacturing strategies to support individualized medicines of the future. The US Food and Drug Administration (FDA) is charged with protecting and promoting the public health by ensuring these products are safe and effective. To that end, FDA's recent approval of a 3D-printed drug product in August 2015 (SPRITAM®) introduced a new chapter in pharmaceutical manufacturing for solid oral dosage form. Although the first approved 3D-printed drug product is a solid oral immediate release product, the majority of 3D printing research for oral delivery has been focused on controlled release, targeting and precise delivery for extremely low dose drugs. Various principles of 3D printing of oral drug products could support production of complex and personalized medicines on demand. However, standardization of some aspects of the 3D printed dosage form and associated manufacturing process may be required to achieve this objective. In this presentation, technical and regulatory considerations for development and control of 3D printed oral products will be discussed for various 3D printing technologies.

DISCLAIMER

This presentation reflects the views of the presenter and should not be construed to represent FDA's views or policies.

531710 Two-Color Photo-Inhibited Systems for Rapid Additive Manufacturing

Abstract:

Additive manufacturing (AM), also known as 3D-printing, is exciting disruptive technology that may upend contemporary manufacturing practices. Production in AM generally proceeds by layer-by-layer addition of materialâ €"typically thermoplastic or photopolymerizable resinâ€"to gradually build up the desired geometry, allowing for the production of highly individualized parts with near limitless complexity. The achievable print speed is still one of the major limitations of the current technologies with layer-by-layer devices only able to achieve vertical print speeds of several millimeters per hour. Recently, a layer-less (or continuous) method for resin based AM, termed continuous liquid interface production, was developed. The method relies on a thin diffusion-controlled oxygen-inhibited dead zone to eliminate adhesion to the projection window during printing, allowing for an order of magnitude increase in the achievable printing speeds when compared with traditional SLA devices.

Photo-inhibitor molecules that inhibit polymerization in the presence of light may be an attractive alternative inhibition mechanism to oxygen-inhibited systems. The advantages of this type of system are that diffusion of oxygen no longer controls the thickness of the dead zone, and the limit of this zone is theoretically much larger. Larger dead zones allow faster polymer reflow under the printing part and, therefore, faster build speeds. This talk will present a novel method using a two-color irradiation scheme to spatially control polymerization to generate dead zones for rapid AM and enable additional functionality over conventional continuous additive manufacturing methods.

531942 Glass-Forming Polymer Networks for Shape-Memory Contact Printing

Abstract:

Current methods for microscale pattern transfer fail to combine high feature resolution, ability to pattern large areas, high throughput and cost effectiveness. Over the past years, several researchers have demonstrated how films of shape-memory polymers can exhibit topographical memory, leading to switchable adhesion or optical properties. We are designing and developing new stimuli-responsive materials to ultimately enable high fidelity pattern transfer by thermal modulation of the elastomeric contact printing process.

Different glass forming polymers that can withstand high elongation and store large amounts of elastic strain energy in compression are under investigation. Two systems will be described: (i) small molecule thiol-ene networks that are softened by addition of dithiol reagents, and (ii) glass forming t-butyl acrylates that are lightly crosslinked. Their shape-memory behavior in tension and in compression will be reported. Furthermore, a custom-built materials testing / contact printing apparatus will be used to set compressive strain between two soft surfaces while measuring stress and contact area. Contact mechanics experiments are modeled and interpreted using finite element analysis to offer deeper insight to the nature of contact and detachment.

523626 Design, Construction and Operation of a Heat Exchanger Test Bed Unique for Leaks Detection and Modeling

Abstract:

Design, Construction and Operation of a Heat Exchanger Test Bed Unique for Leaks Detection and Modeling

Dan Fernandes[a], Daniel Chen[a], Tae Hoon Kim[b]

a-: Department of Chemical Engineering, Lamar University

b-: Department of Industrial Engineering, Lamar University

Key Words-: Heat Exchanger, modelling, corrosion, detection, abnormal situation management

The Texas Commission on Environmental Quality has laid certain guidelines for the control of highly reactive organic volatile compounds (HRVOCs) from the cooling tower heat exchanger system (Rule 115.760) under the Texas Administrative code, Chapter 115 rules in title 30 for compounds. HRVOCs are defined as ethylene, propylene, 1,3-butadiene. HRVOCs are emitted from flares, process vents, cooling towers, and fugitives [1]. Heat exchangers are one of the major sources of fugitive emissions into the cooling water supply. This is largely due to the tube wall perforation /pits development caused by erosion-corrosion on the tube side of the heat exchanger where normally a higher pressure fluid flows. The shapes of the pinholes are irregular or gully in nature. [2] The pits can in single or in groups of several pits close to each other. Coolant and its velocity play an important role in the failure and performance of a heat exchanger. [3]

A unique test bed for the purposes of modelling and leak detection was designed, constructed and operated at Lamar University's Abnormal Situation Management Lab. The shell side fluid is water and the tube side fluid is air or water at a higher pressure than the shell side fluid. The shell is made up of a see-through PVC 6" diameter pipe which is 5'7" long in length. There are four tubes inside the shell made of copper material. One of the tubes has no holes and it is the standard tube for the measurement of air/ water flow. Here modelling of leaks takes place from the tube side fluid which is at higher pressure to the shell side fluid at lower pressure. The other three tubes have holes of 0.0403 inch, 0.0625 inch and 0.0730 inch respectively. There are around 10-12 holes per tube which are close to each other so as to mimic a real-life pitting of the tubes. Fig.1 shows the tube with 0.0625-inch hole size.

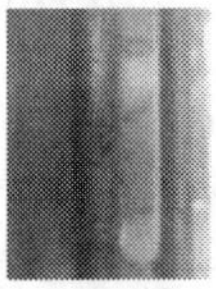

Fig.1: 0.0625-inch hole size mimicking a pit type of corrosion in heat exchanger tube

This test bed is very useful for modelling leaks of different geometries from individual tubes of a heat exchanger which is susceptible to erosion-corrosion. Here tubes of different corrosion geometry (pitting, Stress corrosion cracking) can be studied along with bubble dynamics of the fluid escaping the tubes using high speed camera. The fluid in the tube can be either liquid or gas. Furthermore, novel techniques of detecting leaks in heat exchanger can be studied, we are studying whether leaks can be detected by means of acoustical signature. An imaging sensor such as a digital camera, Fig. 2, will also be used for gas bubble measurements. Computation Fluid Dynamics modelling of leaks in heat exchangers of this test bed will be explored.

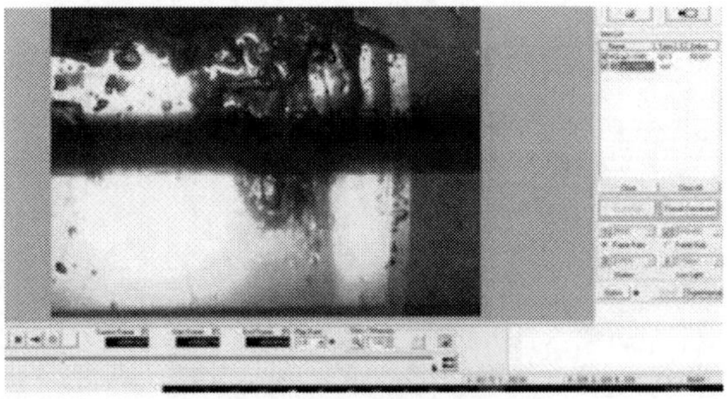

Fig.2. Bubble dynamics of leaks from tube side to shell side using high speed camera @ 5000 fps

References:

1) http://texreg.sos.state.tx.us/public/readtac$ext.TacPage (Texas Administrative Code, Division 2, Cooling water heat exchange systems, Rule 115.764, Monitoring and Testing Requirements)

2) B. Kuznicka "Erosion-Corrosion of heat exchanger tubes" Engineering Failure Analysis, Vol.16,2009,pg 2382-2387

3) Khalil Ranjbar" Effect of flow induced corrosion and erosion on failure of a tubular heat exchanger", Materials and Design, Vol.31,2010, pg.613-619

533167 Design of a Test Rig for the Simulation of Startup Procedures in Main Heat Exchangers of Air Separation Plant

Abstract:

The Kopernikus project â€œSynErgieâ€ investigates an enhancement of the load flexibility of air separation units. Aluminium plate fin heat exchangers (PFHE) are used in these plants because of their high process integration, low production cost and compact design. The main heat exchanger was identified as one of the key components for a dynamic plant operation because startup, shutdown and load change procedures can lead to thermal stress and an increased lifetime consumption.

The transient behavior of temperature profiles within a PFHE can be calculated with satisfying accuracy. Furthermore, a FEM-based model is available to investigate the thermal stress induced by dynamic temperature changes in PFHE. Hence, the reduction of lifetime due to dynamic plant operating scenarios can be estimated and critical operating modes can be identified.

For further development of the model, a PFHE test rig is designed which is representative for extreme operating conditions in air separation units. A test scenario matching the conditions of plant startup after extended downtime in cold condition has been designed. In order to reach a level of thermal stress comparable to industrial sized plants, the PFHE weighing around 1.5 tons is heated up to 50 °C before rapidly being cooled down by a stream of gaseous nitrogen at -173 °C in a cyclic operation scheme.

This test scenario leads to thermal stress that is expected to damage the PFHE within a few weeks of testing. The location of maximum thermal stress in the FEM model matches typical cracks observed in air separation units. The test rig is designed to gain a clearer understanding of the underlying damage mechanism and to evaluate measures for lowering thermal stress in further test series.

537331 Syngas Chemical Looping and Coal Direct Chemical Looping Processes for Hydrogen and Power Production with in-Situ Carbon Capture: Pilot Scale Development and Demonstration

Abstract:

Redox Chemical Looping processes use a metal oxide or metal sulfate to performed reduction and oxidation reactions with a carbonaceous fuel and air/steam, respectively. Redox chemical looping systems are advantageous in their ability to produce high purity products undiluted with nitrogen present in air, eliminating the need for energy intensive components such as a cryogenic air separation unit or an acid gas removal device. The Ohio State University (OSU) has advanced the redox chemical looping concept in the development of two, 250 kW_{th} pilot-scale demonstration plants for coal combustion to produce heat for electricity generation, the coal direct chemical looping (CDCL) process, and for gaseous fuel conversion for producing high purity H_2, the syngas chemical looping (SCL) process. In each of these processes, a countercurrent moving bed reducer reactor is used for full conversion of the carbonaceous fuels to a gas stream consisting predominantly of CO_2 and H_2O. A fluidized bed combustor reactor is used to regenerate the oxygen carriers to their original oxidation state. In the SCL process, a third countercurrent moving bed oxidizer reactor is used to produce high purity H_2 from steam. The countercurrent moving bed reducer reactor design in the CDCL and SCL processes ensure nearly full fuel conversion to CO_2 with minimal solid circulation requirement and have the capability of producing high purity H_2. The SCL pilot plant was constructed and tested at the National Carbon Capture Center for processing coal-derived syngas to H_2. The CDCL pilot plant was constructed with ongoing operations at the Babcock and Wilcox Research Center for coal combustion with carbon capture. The present paper discusses the development strategy in the design of both pilot plants, reviews the startup and operation challenges observed during the plant commissioning, and summarizes the demonstration results to date. Additional discussion is presented on the comparison the demonstration results with the theoretically expected performance. Overall, over 1,500 hours of operational experience were gained from these pilot plants.

541196 Electrochemical Production of Ammonia from Nitrogen and Water for Electrical Energy Storage

Abstract:

Electrical energy storage is much needed to increase usage of wind and solar power. Ammonia is an attractive energy carrier from points of views of energy density, transport, and storage. A new electrochemical conversion process will be introduced to produce ammonia from air and water at low temperature and low pressure.

523832 Thermal Energy Storage (TES) with Silica Gel Regenerated at Low Temperature

Abstract:

The energy system around the world is facing technology transformation. The deployment of renewable energy generation and the development of new clean technologies offer the opportunities to diversify energy supply, improve energy independence and reduce carbon dioxide (CO_2) emission [1][2]. Although it is environmentally beneficial to use renewable energy in the electricity production, heating purpose and transportation, one of the main difficulties faced by the energy industry is the absence of economic and reliable energy storage solutions [3]. Thermal energy storage (TES) provides a solution to store energy generated from different types of energy sources (traditional or renewable) and correct for the mismatch between the energy supply and demand. TES through water vapor adsorption process has attracted increasing interest for its thermal applications such as space heating and cooling.

Various kind of materials have been investigated in search of adsorbents with high energy density for TES. Most of the energy density data reported were either theoretical data or experimental results obtained at high material regeneration/desorption temperature (e.g. 135 – 250 C) [4]. However, experimental energy density of the adsorbent varies if the design and the construction material of the system and/or the operating conditions change. Therefore, the experimental energy density could be much lower than the theoretical energy density. Moreover, for most of the residential renewable energy heating sources, such as solar water heating system [5] or building-integrated photovoltaic/thermal (BIPV/T) systems [6], the temperature of dry air that the system could provide for material regeneration is less than 80 C.

In our previous study [7], silica gel has been identified as a promising adsorbent candidate for thermal energy storage applications. In this study, a commercial silica gel material was used to examine the effects of relative humidity (RH) on the material behaviour and its experimental energy density at low regeneration temperature (50 – 80 C). The experimental results showed that an energy density of 120 kWh/m^3 could be obtained under cyclic operating conditions with low RH and low regeneration temperature.

References

[1] B. Moreno and A. J. López, "The effect of renewable energy on employment. The case of Asturias (Spain)," *Renew. Sustain. Energy Rev.*, vol. 12, no. 3, pp. 732–751, 2008.

[2] P. del Rio and M. Burguillo, "Assessing the impact of renewable energy deployment on local sustainability: Towards a theoretical framework," *Renew. Sustain. Energy Rev.*, vol. 12, no. 5, pp. 1325–1344, 2008.

[3] U. Lehr, J. Nitsch, M. Kratzat, C. Lutz, and D. Edler, "Renewable energy and employment in Germany," *Energy Policy*, vol. 36, no. 1, pp. 108–117, 2008.

[4] F. Fischer, E. Lävemann, A. Krönauer, and A. Hauer, "Open adsorption systems for thermal energy storage applications," in *BIOWKK Workshop Dortmund 2012*, 2012.

[5] R. Shukla, K. Sumathy, P. Erickson, and J. Gong, "Recent advances in the solar water heating systems: A review," *Renew. Sustain. Energy Rev.*, vol. 19, pp. 173–190, 2013.

[6] T. Yang and A. K. Athienitis, "A review of research and developments of building-integrated photovoltaic/thermal (BIPV/T) systems," *Renew. Sustain. Energy Rev.*, vol. 66, pp. 886–912, 2016.

[7] Y. Hua, A. Godin, T. Alenko, B. Ugura, and F. H. Tezel, "Adsorbent Screening for Thermal Energy Storage Application," in *14th INTERNATIONAL CONFERENCE ON ENERGY STORAGE*, 2018.

537142 Efficient Hydrogen Production from Solar Thermal Energy Via High Temperature Water Electrolysis

Abstract:

With the rapid growth in world population and increasing demand for life quality, human beings are continuously exerting immense pressure on our global energy resources. However, even with development in renewable energy application, the world energy consumption is still largely dependent on fossil fuels, which have brought environmental issues such as greenhouse gas emission and might be deprived in a foreseeable future. Switching energy resources to renewable energy will be crucial to the global energy challenge and solar energy is the promising because of its abundance and wide availability at most locations in the world. Due to the limitations on land area that can be dedicated to solar energy harnessing and the uncontrollable intermittencies in the solar energy supply, technologies are highly desired to fully utilize the solar energy that can be harnessed and convert or store it in the form of electricity, chemicals, etc. With its high energy density and zero greenhouse gas emission, hydrogen is the key energy carrier in a sustainable future. Hydrogen production from solar energy has been studied by many researchers by either thermochemical or electrolysis method.

Here we introduce a process design strategy for the production of hydrogen by high temperature water electrolysis using concentrated solar thermal energy. High temperature water electrolysis is realized by using solid oxide electrolysis cell (SOEC). SOEC requires both electricity and heat input to enable the energy balance of the water splitting process. We utilize solar thermal power production cycle to supply the electricity the SOEC requires. Meanwhile, water is heated by concentrated solar energy to reach the SOEC operation temperature before electrolysis. The power generating and hydrogen production processes are integrated, allowing electricity flow and heat transfer between the two cycles. Process simulations for the proposed integration are performed in an integrated Matlab and Aspen Plus platform.

Considering the electricity and power input as well as the overpotential of the SOEC, different operation modes of SOEC, including thermoneutral, endothermic and exothermic operations, could largely affect the efficiency of hydrogen production. Therefore, different operation modes of SOEC are investigated and various process designs are examined accordingly to figure out the most efficient process under various circumstances. In this way the most promising solar thermal hydrogen production process could be designed and flexibility of operation conditions could also be identified.

539306 Monitoring and Classification System for Water Recycling

Abstract:

Minimizing water consumption and disposal has huge benefits in terms of safety and cost savings. However, to maximize the amount of water that is recycled at a site generating multiple complex waste streams, an intelligent water monitoring system is needed to ensure that water from various sources is processed appropriately. Mainstream Engineering has developed a chemometric water monitoring system (CWMS) to fully characterize water prior to recycle or disposal. This technology relies on statistical methods that relate data to properties of interest. Specifically, we used a combination of several rapid, reagentless sensors to characterize the water according to several classes.

We demonstrated the CWMS for a variety of wastewater sources including laundry, kitchen, and latrine. We then validated our models with additional samples from other water sources such as wastewater treatment facilities to demonstrate the robustness of the system. After specifying all of the components, we designed a prototype system. Principal component analysis was used to classify water and draw distinctions between surface non-potable water, greywater and blackwater. We were also able to eliminate the need for dedicated COD, BOD_5, turbidity and chlorine sensors by processing spectroscopic data with chemometrics.

522532 Water Recovery and Reuse in Soluble Coffee Production Using a Dynamic Membrane Process

Abstract:

Through a grant from the U. S. Environmental Protection Agency, Pollution Prevention program, we have partnered with NestlÃ© USA to explore the use of novel membrane technology to improve the manufacturing efficiency of soluble coffee manufacturing operations. Our case study at their Freehold, NJ factory, evaluates water recovery methods for a spray dried and freeze dried coffee manufacturing plant. The particular focus of this paper is on a dynamic / shear-enhanced vibratory field membrane method to recover water for plant operations, thus reducing operating costs and environmental footprint. This research helps address a major issue in the food manufacturing sector â€" the efficient use of water and waste reduction.

The production of spray dried and freeze dried coffee involves many highly complex processes that rely on large amounts of water. These operations include process water for cooling, water to be heated to produce steam, water for process equipment operations, water for intermediate production steps, and water for cleaning and sterilization. When utility generation is included, such as operation of cooling towers, water consumption can be significant.

Many of the problems associated with effective recovery, involve the separation of dissolved, colloidal, and particulate matter, to obtain water of suitable quality for reuse. Suspended and colloidal matter is one of the biggest causes of fouling and performance degradation in operating these water recovery systems. This causes additional expense, and in many cases industry abandons this approach, since it will not produce sufficient quantity and quality of water for extended periods of time. One potential solution that is described in this paper, is a vibratory-field membrane process for food/biomass waste processing.

Our study describes the benefits and limitations of this technology, and how laboratory investigations help guide industrial use and scale-up considerations. Various membrane types were evaluated to help inform decision makers based on the reuse goals. The study evaluated the overall savings and reduction in life cycle emissions for the recovery of water from a waste stream, for use in cooling tower operation. The assessment included savings, both economically and environmentally, in the reduction in fresh water intake, waste pretreatment, and final disposal. These are compared to the increased cost and operational factors of a new recovery system. The case study found that the greatest cost and environmental impact reductions were associated with the reduction in wastewater discharge. Long-term economics were more highly influenced based on equipment size. This case study is applicable to the food and beverage manufacturing sector.

525339 How Active Is Too Active?Â a Catalyst Selection Study

Abstract:

When selecting a catalyst formulation as part of a new reactor build, the process can be designed around the catalyst such that nearly any commercially-viable catalyst can be used. For these systems, higher activity catalysts generally give rise to higher product rates or reduced capital investment. However, most of industryâ€™s active processes are not newly built, but have been in service for many years (if not decades). For these systems, higher activity catalysts can also enable higher product rates, *but typically only to the extent constrained by the original process design*. Indeed, many if not most of todayâ€™s active systems are now multiply constrained, such that it is not always true that increasingly higher activity catalysts will yield better performance.

As one example, many fixed-bed, vapor-solid reactor systems are constrained by heat transfer. For these systems, a circulating coolant absorbs the heat of reaction thereby controlling the temperature rise and making possible higher product rates. In this context, there is generally a maximum achievable product rate corresponding to the maximum heat removal capability of the system. This maximum product rate is itself a function of the catalyst (in particular, the catalyst activity and selectivity).

In choosing from next-generation, high activity catalysts for a system constrained by heat transfer, the impact of varying activity and selectivity must be balanced, taking into account also the free variables for optimization and their bounds. Here, we utilize detailed reactor simulations to investigate the catalyst selection problem for one such industrial system. Our steady-state simulations incorporate mass and energy balances and reaction kinetics while predicting temperatures and concentrations that vary in two spatial dimensions. We begin by demonstrating how such simulations, when informed by laboratory experiments, can predict the heat transfer-constrained operation as observed in the plant. Then, we utilize the tuned reactor model to explore how process optimization accounting for a catalyst change (e.g., changes to the base reaction rates, activation energies, and partial reaction orders) leads to varying and sometimes counterintuitive changes in reactor performance. To finish, we develop simplified analytical models from first principles and compare their predictions to those generated using the full reactor model. These simplified models can be utilized as a coarse guide for catalyst selection in the case that more detailed and highly tuned reactor models are not available.

530554 Glycol Loss Minimization for a Natural Gas Dehydration Plant Under Upset Conditions

Abstract:

Low-temperature separation with Mono Ethylene Glycol (MEG) injection process is a common dehydration technique for natural gas processing, which needs to maintain a hydrocarbon dew point. However, dehydration systems based on glycol frequently result in a significant loss of glycol during plant upset conditions such as rapid flow fluctuation, refrigeration system failure and inadequate control of stripper column [1]. These losses occur due to carryover with gas, hydrocarbon and vaporization from stripper column overhead. For this reason, it is important to develop process operation methodologies for reducing glycol losses and thereby reducing emissions.

In this research, steady-state and dynamic models are developed using the Aspen Plus simulation framework. A parametric analysis of inlet flow fluctuation, temperature fluctuation of the low-temperature separator (LTS), stripper column overhead temperature fluctuation and reboiler temperature fluctuation were conducted under steady-state conditions. Simulation results indicated that that most of the glycol is lost via vaporization through the stripper column overhead and some glycols are carried over with gas and hydrocarbon from LTS. Dynamic simulations were utilized to develop a new control strategy for optimum column operation along with an improved stripper model. It is shown that the simulation results can be used to study plant upset conditions, minimize glycol loss, improve product gas specifications and reduce plant operating costs. In particular, it is shown that the improved design has the potential to reduce glycol losses by over 70%, lower emissions and provide plant stability.

References

1. Haque, M. E.; Ethylene Glycol Regeneration Plan: A Systematic Approach to Troubleshoot the Common Problems. *Journal of Chemical Engineering*, *IEB, 27, No. 1, 21-26, 2012.*

Hydrodeoxygenation of Karanja Oil for the Production of Green Diesel: Process Design with Heat Integration and Economic Analysis

Swarnalatha Mailaram and Sunil Kumar Maity,
IIT Hyderabad, Hyderabad, India

Introduction

Major concern on environmental effects and depletion of fossil fuels led to search for the alternative resources for energy. Biomass is the best possible alternate source in terms of its availability in nature and it also has the potential to transform to all forms of energy including transportation liquid fuel which is the leading energy consuming sector with ~27% share globally [1]. Much research and technological development has been done so far in the production of biofuels ex: bioethanol, biobutanol, biodiesel etc. Biomass derived oxygenated biofuels suffer with low mileage because of the presence of oxygen and these biofuels are not compatible with existing infrastructure. Transesterification is one of the technology producing bio diesel which is limited to blending with petro-diesel due to the presence of oxygen which lowers the fuel mileage. Therefore, there is a need of hydrocarbon fuels similar to petro-diesel, which is compatible with existing combustion engine. Inedible vegetable oil consisting of long hydrocarbon chain with low oxygen content and lesser complexity of chemical structure compared to lignocellulosic biomass is the promising biomass to produce transportation fuels. Karanja oil production potential is more in south east Asia with 78.6wt% C_{18}, 11.3wt% C_{16}, 3.7wt % C_{20}, 6.4wt% C_{22} fatty acid composition [2]. Hence, Karanja oil is chosen as the feedstock.

Hydrodeoxygenation process

Hydrodeoxygenation (HDO) is the route which removes oxygen content in the triglyceride in the form of water to form hydrocarbon diesel range fuel called as green diesel. Green diesel has higher energy density (44 MJ/kg), high cetane number (70-90) and can directly be used in existing combustion engines [3][4]. UOP has recently developed a two-step ecofining process aiming to produce green diesel by HDO [5]. Studies on various oils by HDO process has been performed by researchers. HDO is generally carried out at high hydrogen pressure (25–150 bar) over supported metal catalyst in the temperature range of 523 K to 773 K [6]. Generally, HDO of Karanja oil can be envisaged through two different routes: (i) direct HDO and (ii) two-step HDO (hydrolysis of vegetable oil followed by HDO of fatty acid). In the Direct HDO, hydrogenation of unsaturated triglyceride in Karanja oil gets saturated triglyceride and undergo HDO to form diesel range hydrocarbons (C_{18}-C_{22}) termed as green diesel, water, traces of CO_2 with propane as a co-product. In the second route, Karanja oil is first hydrolyzed with water to form fatty acids and glycerol (co-product). These fatty acids then undergo HDO forming green diesel, water and traces of CO_2 [3][4][7].

Process design with heat integration

In the present work, process design for both the routes were performed using Aspen Plus software for various capacities of plant ranging from 0.01 to 0.15MMT of Karanja oil /annum. Schematic process for direct HDO process is shown in Fig. 1. In direct HDO, karanja oil and hydrogen enter the HDO reactor operating at 623K, 100bar. Excess hydrogen is removed in a separator following reactor. The product stream containing green diesel, water, CO_2, propane is sent to downstream processing. Water, green diesel and gases (propane and CO_2) are separated and gases are further sent to absorber and stripping unit in which monoethanolamine (MEA) is used as a solvent to separate CO_2 from propane and propane is the revenue generating co-product. As shown in Fig. 2., In two step HDO, karanja oil is hydrolysed with water in hydrolysis reactor at 533K and 56bar to form fatty acids and glycerol. Glycerol and water is separated and sent to the glycerol refining section to obtain 99.99wt% glycerol which is the co-product in this process. Fatty acids formed are sent to the HDO reactor to form where green diesel, water and traces of CO_2 are formed. Green diesel is separated from gases and water. Heat integration for both the processes has been done to reduce the utility requirement and optimize the heat transfer within the plant. Pinch analysis has been chosen to obtain the minimum energy consumption in the plant and design heat exchange network for both the processes. Resulting grand composite curve (GCC) by pinch methodology, was useful to obtain the minimum hot and cold utility requirement and heat recovery within the system. Based on the possibility of heat recovery in the process and pinch analysis criteria, heat exchange network has been designed. This helped in redesigning the processes with the heat exchange network system.

Economic analysis

The economic evaluation has been carried out for different plant capacities ranging from 0.01 to 0.15MMT/annum for both direct HDO and two-step HDO process. All possible cost contributing factors were considered to obtain capital and operating expenses for both the processes. Capital investment for both the processes is practically identical. Impact of plant capacity on both processes is huge. The production cost was reduced with increasing plant capacity. The reduction in the production cost was, however, negligibly small with increasing plant capacity beyond 0.12 MMT/annum. Hence, 0.12 MMT/annum has been considered as the optimum plant capacity. The production cost of green diesel was estimated as 0.84 $/kg and 0.798$/kg for direct and two-step HDO respectively for the plant capacity of 0.12 MMT Karanja oil per annum. Also, contribution of various costs to production cost has been studied. Feedstock alone is contributing around 73-77% to the production cost which is very high. Contribution of co-product credits has significantly reduced the production cost of green diesel in two-step process. Utility consumption reduction with the help of heat recovery in the process has shared 1.8-4.18% which was ~35% contribution without heat recovery. Price of co-products has also a significant impact on the production cost. Production cost was decreased by 0.5% and 1.6% for 10% increase in the price of co-products for direct and two-step HDO respectively. The discounted cash flow analysis was further carried out to evaluate the economic performance and determine the minimum selling price of green diesel. The price of green diesel excluding taxes was found to be 1.186 $/kg and 1.21 $/kg for an acceptable rate of return of 8.5% for direct and two-step HDO respectively. The green diesel can be considered as economically competitive with the retail price of the petro-diesel.

Summary

Conceptual processes of direct and twostep HDO is designed using Karanja oil as feedstock. Maximum energy recovery using pinch methodology has been done. Heat exchange network has been designed for both the processes. Heat recovery within the system with the use of pinch analysis has reduced the utility consumption in the process. Thereby, reducing the production cost of green diesel significantly. Economic analysis for modified processes with heat exchange network was evaluated. Capital, operating expenses, production cost of green diesel is estimated. Feedstock cost plays key role in reducing the cost of the diesel. If reduced, drastic decline in the cost can be observed. Discounted cash flow analysis was used to find the minimum selling price of green diesel.

References

1. Annual Energy Outlook. U.S. Energy Information Administration. Available at: http://www.eia.gov/forecasts/aeo/index.cfm: 2015.
2. P. Kumar, S.R. Yenumala, S.K. Maity, D. Shee, Kinetics of hydrodeoxygenation of stearic acid using supported nickel catalysts: Effects of supports, Appl. Catal. A Gen. 471 (2014) 28–38.
3. S.R. Yenumala, S.K. Maity, D. Shee, Reaction mechanism and kinetic modeling for the hydrodeoxygenation of triglycerides over alumina supported nickel catalyst, React. Kinet. Mech. Catal. 120 (2017) 109–128.
4. Kalnes TN, Marker T, Shonnard DR, Koers KP. Green diesel production by hydrorefining renewable feedstocks. Biofuels Technol 2008:7–11.
5. UOP LLC, A Honeywell Company. Honeywell Green Jet Fuel: Sustainable fuels for cleaner skies. 2011. Available at: https://www.uop.com/?document=honeywell-green-jet- fuel-brochure&download=1
6. Glisic SB, Pajnik JM, Orlovic´ AM. Process and techno-economic analysis of green diesel production from waste vegetable oil and the comparison with ester type biodiesel production. Appl Energy 2016;170:176–85.
7. S.R. Yenumala, S.K. Maity, D. Shee, Hydrodeoxygenation of karanja oil over supported nickel catalysts : influence of support and nickel loading, Catal. Sci. Technol. 6 (2016) 3156–3165.

Figures

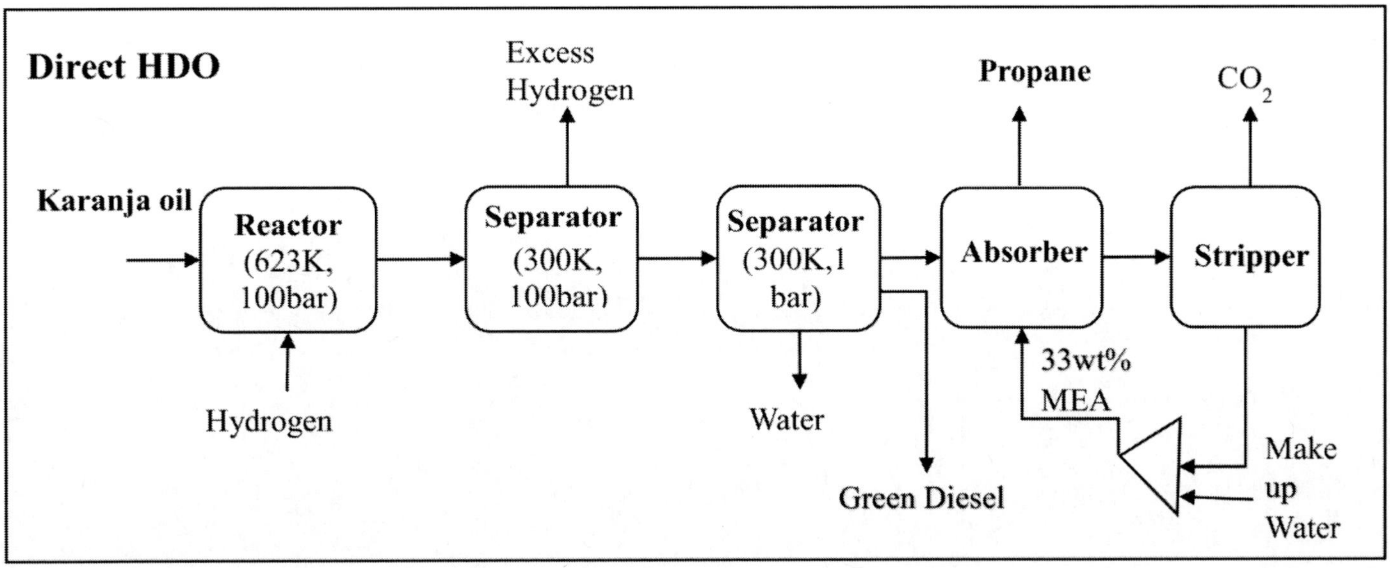

Fig. 1. Schematic process flowsheet of Direct HDO process.

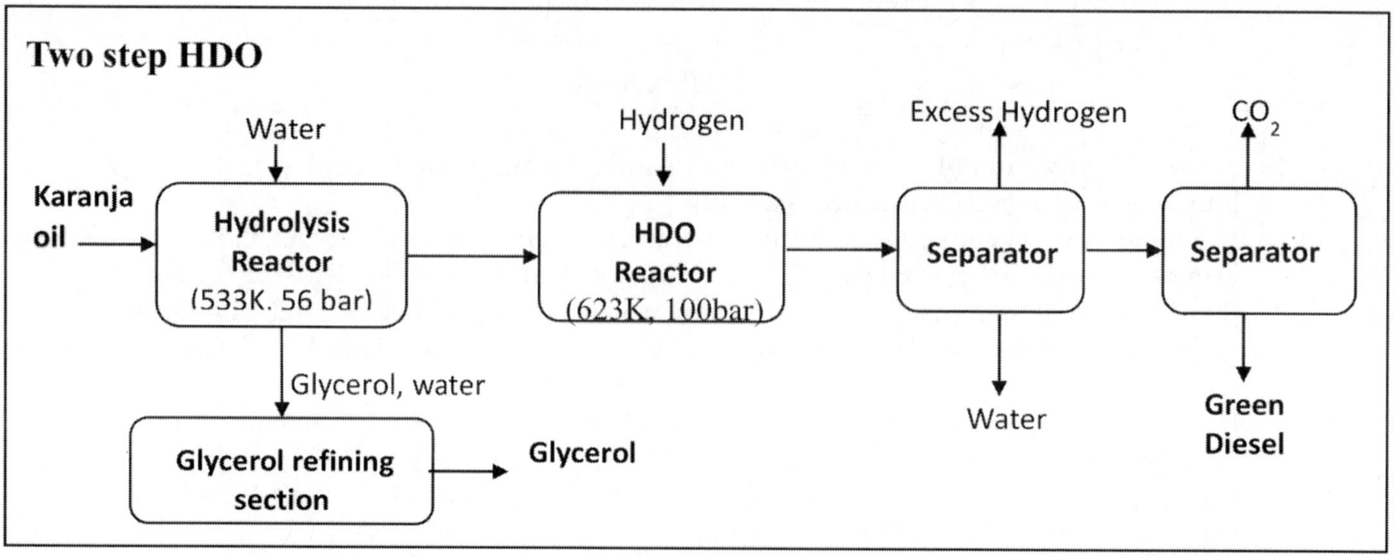

Fig. 2. Schematic process flowsheet of Twostep HDO process.

531870 An Integrated Process for Desulfurization (HDS-Extractive-Oxidative)

Abstract:

An integrated process for Desulfurization (HDS-Extractive-Oxidative)

Mokhtarian, M.R. Dehghani, F. Banisharif

Chemical Engineering Department, Iran University of Science and Technology, Narmak, Tehran, Iran

Abstract:

Available sulfur compounds in the fuels such as thiophens are considered as one of the main source of air pollution. They converted to SOx and released to the atmosphere as a harmful chemical for human health. Hence, removal fuel desulfurization is mandatory from environmental point of view as well as public health issues. Nowadays, in many countries, stringent regulations are being enacted to reduce the sulfur content to lower than 10 ppm for both diesel and gas oil. During recent years many attentions have been paid to extractive-oxidative desulfurization (EDS-ODS) while economic evaluations show that they are not economically feasible compared to the hydro desulfurization (HDS) process. On the other hand, HDS process is considered as an energy insensitive process and needs to be operated at High pressure and temperatures. To present an economic process, in this work an integrated process of HDS and EDS-ODS were modeled. The HDS section was modeled based on operational data from Tehran refinery and extractive-catalytic oxidative desulfurization (ECODS) part were modeled based on our novel pilot experimental data. Gas oil as the feedstock with total sulfur content of 22600 ppmw was fed to the process. Optimal conditions were obtained using the response surface methodology and box-behnken design (BBD), based on experimental design results, hydro desulfurization from the 22600 ppmw to 1930 ppmw level was done under the following conditions: hydrodesulfurization temperature: 311 â„ƒ, hydrodesulfurization pressure: 24 bar and hydrogen purity of 35 wt%. Then gas oil with the total sulfur content of 1930 ppmw was entered into the oxidative-extractive desulfurization unit. The effects of different operating parameters such as temperature, oxidant to sulfur molar ratio and the type of solvent in the extraction stage were investigated. The best performance was obtained at oxidant to sulfur molar ratio of 4:1 and acid to oxidant volumetric ratio of 1:1. It must be mentioned that hydrogen peroxide (30%) and acetic acid (95%) were used as oxidant. Finally, a liquid-liquid extraction process was utilized for deep desulfurization step. Acetonitrile used as solvent at 1:1 ratio towards gas oil. It was found that the proposed process has lower costs and also is inherently safer than that of conventional HDS, ODS and EDS.

Keywords: Optimization, Hydrodesulfurization, Extractive- Catalytic oxidative desulfurization, Ultra-low Sulfur Fuel

540773 2018 Outlook for Energy: A View to 2040

Abstract:

The Outlook for Energy is ExxonMobil's global view of energy demand and supply through 2040. It also highlights the dual challenge of providing affordable energy to support prosperity while reducing environmental impacts – including the risks of climate change. We use the Outlook to help inform our long-term business strategies and investment plans.
The Outlook anticipates significant changes through 2040 across the world to boost living standards, reshape the use of energy, broaden access to abundant energy supplies, and accelerate decarbonization of the world's energy system to address the risks of climate change.

- Global energy needs will rise about 25 percent over the period to 2040
- Electricity demand will nearly double in non-OECD countries
- Worldwide electricity from solar and wind will increase about 400 percent
- Natural gas will expand its role, led by growth for electricity generation
- Growth in oil demand will be driven by commercial transportation and chemical needs
- Efficiency gains and growing use of less-carbon intensive energy sources will contribute to a nearly 45 percent decline in the carbon intensity of global GDP

Energy is fundamental to modern life, and as the world's population approaches 9 billion people in 2040, we are challenged to improve living standards everywhere. Understanding the factors that drive the world's energy needs – and likely choices to meet those needs – is the mission of the Outlook. By sharing the Outlook with the public, we hope to broaden that understanding. Energy matters to everyone, and we all play a role in shaping its future.

541543 Energy Decarbonisation Scenarios

Abstract:

With the energy sector being associated with 2/3 of the annual greenhouse gas emissions, decarbonizing the energy is a key step to reduce their level. The National Determined Contributions (NDC's) announced so far by the participating countries do not meet the objectives of the Paris-climate agreement, as far as reaching a peak in emissions in the near future or decreasing to a level that is compatible with the stated long-term targets.

Using different scenarios for energy demand and supply, options for energy sector decarbonisation are evaluated. Meeting the 2-degree target requires a major transformation in the power sector, efficiency improvements – specially in the buildings and in the industry, and a revolution in transport towards fuel and modal shifts. With a significant increase in renewable energy, system integration and optimization will also be key in reducing the cost of the energy transition. Meeting more ambitious climate goals ("well below 2 degrees") will push further the technological innovation and implementation requirements throughout the entire energy value chain. A scenario aiming at zero net-emissions by 2060 is presented, along with a comparison to the 2-degree case.

542625 Fundamental Research Needs to Advance Energy Technologies

Abstract:

The U. S. Department of Energy (DOE), Office of Science, Office of Basic Energy Sciences (BES) supports fundamental research in chemical and materials sciences to provide the foundations for new energy technologies and to support DOE missions in energy, environment, and national security. This presentation will discuss opportunities for fundamental research to impact DOE's energy mission "to catalyze the timely, material, and efficient transformation of the nation's energy system and secure U.S. leadership in energy technologies" with a focus on the way we generate, store and use energy nationally. I will provide an overview of BES strategic planning over the past decade that identified priority research directions for advancing energy applications, highlight key scientific advances in these areas, and discuss some future opportunities for modern science, particularly chemical sciences, to accelerate the transformation of the U. S. energy portfolio.

540924 Nano- and Microfabricated Hydrogels for Regenerative Engineering

Abstract:

Engineered materials that integrate advances in polymer chemistry, nanotechnology, and biological sciences have the potential to create powerful medical therapies. Our group aims to engineer tissue regenerative therapies using water-containing polymer networks, called hydrogels, that can regulate cell behavior. Specifically, we have developed photocrosslinkable hybrid hydrogels that combine natural biomolecules with nanoparticles to regulate the chemical, biological, mechanical and electrical properties of gels. These functional scaffolds induce the differentiation of stem cells to desired cell types and direct the formation of vascularized heart or bone tissues. Since tissue function is highly dependent on architecture, we have also used microfabrication methods, such as microfluidics, photolithography, bioprinting, and molding, to regulate the architecture of these materials. We have employed these strategies to generate miniaturized tissues. To create tissue complexity, we have also developed directed assembly techniques to compile small tissue modules into larger constructs. It is anticipated that such approaches will lead to the development of next-generation regenerative therapeutics and biomedical devices.

Commercial Scale Manufacturing of a Pharmaceutical Product Using Powder-Liquid 3D Printing Technology

Timothy S. Tracy, PhD, Aprecia Pharmaceuticals, Blue Ash, OH

Abstract

Three dimensional printing (3DP) opens the possibility to manufacture pharmaceuticals differently from today and provides the opportunity to create novel dosage forms for patients. Despite significant advances in the field, an obstacle to commercialization of 3DP pharmaceuticals has been the difficulty in being able to manufacture at commercial scale. Recently, utilizing powder-liquid 3DP technology a pharmaceutical (levetiracetam, Spritam™) has been developed through the 505(b)(2) regulatory pathway at commercial quantities and has received FDA approval for manufacture and sale. In addition, the product is a novel dosage form that utilizes fast-melt technology (ZipDose™) that provides for rapid dispersion in the mouth (< 10 seconds) with only a sip of liquid, while containing doses of levetiracetam up to 1000mg. Products that rapidly disperse in the mouth are particularly useful for patients with dysphagia or difficulty swallowing, particularly of large tablets. This rapid dispersion at high doses was possible because of the selective porosity and hardness of the dosage form created by the 3DP layering process. To produce this unique dosage form, processes were developed to assure consistent and precise jetting of print fluid and careful control of powder deposition across up to 30 layers at commercial production speed to produce tablets that were highly accurate and reproducible with respect to dose of active drug. The regulatory approval and commercialization of this product demonstrated that unique pharmaceutical dosage forms could be created through 3DP technology that helped address unmet needs.

540925 Closing the Circle on Design, Hardware, and Materials for Additive Manufacturing

Abstract:

3D printing is a subset of additive manufacturing (AM) and represents an evolving set of tools and capabilities to prototype, and in some cases produce, new products in much faster iteration cycles than traditional subtractive processes. For 3D printing to reach its full potential as a viable large-scale manufacturing option, new advancements are still required across the three major segments of the industry â€“ design, hardware, and materials â€“ and to accelerate these advancements, highly collaborative efforts between the segments will provide 3D printing users with the quickest and best solutions for their production needs. The presentation will highlight the current state-of-the-art in materials for 3D printing, known existing property gaps, Dowâ€™s vision for the future development and performance of 3D printable materials, and how the broader industry can collaborate to shorten the path to wider scale adoption of these tools in 21st century manufacturing.

541541 Additive Manufacturing and Architected Materials

Abstract:

Material properties are governed by the chemical composition and spatial arrangement of constituent elements at multiple length-scales. This fundamentally limits material properties with respect to each other creating trade-offs when selecting materials for specific applications. For example, strength and density are inherently linked so that, in general, the more dense the material, the stronger it is in bulk form. We are combining advanced microstructural design, using flexure and screw theory as well as topology optimization, with advanced additive micro and nanomanufacturing techniques to create new material systems with previously unachievable property combinations â€“ mechanical metamaterials. The performance of these materials is fundamentally controlled by geometry at multiple length-scales rather than chemical composition. We have demonstrated designer properties of these mechanical metamaterials in polymers, metals, ceramics and combinations thereof. Properties include ultra-stiff lightweight materials, negative stiffness, and negative thermal expansion. Our additive micro-manufacturing techniques include Projection Microstereolithography (PmSL), Direct Ink Writing (DIW), and Electrophoretic Deposition (EPD). These tools are capable of generating the designed structures which are highly three-dimensional micro- and nano-scale architectures with multiple constituent materials in the same structure.

540113 Using Team Building to Focus the Intersection of Diversity and Inclusion with Process Safety Practice

Abstract:

Within the past decade or so, process safety practice has undergone an important transition. While the importance of process safety has a long history of practice by many organizations, the profession of chemical engineering began to formalize such practices with the development of Risk Based Process Safety (RBPS) by the Center for Chemical Process Safety (CCPS) of the American Institute of Chemical Engineers (AIChE). RBPS has evolved over time, but its importance was recognized by OSHA early in the process, and OSHA included aspects of RBPS in its Process Safety Management (PSM) requirements. One of the important principles developed and practiced as a part of RBPS is that clear, relevant communication and training is critical to the effective practice of process safety. The importance of Diversity and Inclusion (D&I) is broadly recognized as important in our society, and in its most basic sense, diversity (bringing together people with different backgrounds) and inclusion (all those involved participating on an equal footing) are reflected in many aspects of RBPS. This paper discusses the intersection of D&I with process safety and particularly how training students to work effectively in teams supports these practices.

535890 Developing a Mindset of Safety in Students

Abstract:

Our process safety course â€œSafety, Health, and Loss Preventionâ€, which is an elective populated primarily by seniors, provides an in-depth look at specific elements of safety, some of which are also considered in the required courses of the curriculum. The overarching goals of the course are to provide students with a base level of knowledge of the safety regulatory structure in the US, industrial hygiene, and techniques for safer process design, and the ability to use appropriate models to quantify health and economic effects, toxicant dispersion, and damage due to fire and explosions. This is accomplished through a variety of techniques including in-class modeling exercises, expert guest speakers, and student led presentations as well as traditional lectures and homework assignments. The assessment of their understanding is based on several quizzes, exams, and their presentation.

One of our personal goals is for the students who complete a chemical engineering degree in our program to leave with a keen sense of the importance of safety, especially in the process industry. Our department works to develop that mindset of safety by incorporating small units relating to safety in several of our core courses, including a specific safety unit with activities in the Unit Operations Laboratory course, emphasis on safe behavior in all laboratory courses, and requiring EH&S in their senior design project. These units and activities provide a foundation on which they will build either in their careers or in our Process Safety, Health, and Loss Prevention (aka Safety) course. By including safety information throughout the curriculum, the students are reminded on a regular basis that safety must be a priority.

In this presentation, we will discuss our approach to this course as well as the departmental approach to incorporating safety throughout the curriculum.

527167 Process Safety Instructional Enhancements Implemented in a Two-Semester Chemical Process Design Course

Abstract:

An ABET review of our chemical engineering program in the first year of the authorâ€™s teaching appointment at Texas A&M University-Kingsville (TAMUK) spurred a comment regarding the need for greater process safety instruction in our curriculum. Our department faculty indicated this instruction would best fit into both the unit operations laboratory two-course sequence and the chemical process design two-course sequence. The author, who is the instructor for the chemical process design course sequence, has, over the last two years, incrementally developed and presented new units on chemical process safety for both of the design courses. The emphasis of the new units has been on four different areas. The first of these focus areas is identifying and understanding reactivity, flammability, and toxicity hazards of the chemicals studied in various course design problems, including the group-focused senior design project in the second course. The second focus area is risk-based process safety management systems commonly implemented by refinery and chemical process corporations. The third focus area is on review and analysis of actual chemical accidents, such as through event reconstruction videos provided by the Chemical Safety Board. The fourth focus area is on understanding and creating HAZOPs tables for limited portions of the process each group is developing for their senior design project. Students are evaluated on their understanding of this material using quizzes in the second design course, as well as on the extent to which these principles are included in their senior project chemical process design. The author uses material obtained at the CCPS Faculty Workshop on Chemical Process Safety from June 2016, as well as material from Crowley and Louvarâ€™s *Chemical Process Safety* text and Turtonâ€™s *Analysis, Synthesis and Design of Chemical Processes* text, to develop these instructional units. This paper focuses on presenting the methods used for instruction and the successes and challenges indicated by student work in grasping this course material over the prior two years of chemical process design courses at TAMUK.

537514 Integration of Process Safety Experience in Research Project into Undergraduate Process Design Courses

Abstract:

The undergraduate, senior process design course at The Ohio State University comprises a series of two 4-credit courses, namely: Chemical Process Principles: Economics and Strategy and Chemical Engineering Design and Development. The first course (Process Principles) introduces and teaches senior chemical engineering undergraduate students a core set of process design principles, including process safety evaluation, economic evaluation, optimization theories and heat integration. The second course (Design and development) exposes students to managing, working and communicating on a real world (industry/academia sponsored) open-ended chemical engineering project wherein they apply the skills learned from the first course towards completing the project. This two-tiered structure gives students an in-depth learning opportunity to learn skills in one course and reinforce them by application to a real-world project in the second.

This talk will focus on the process safety facet of the two courses. Specifically, over the last two years, the presenting author and the instruction team have restructured the process safety lessons taught in the first course (skill building) to revolve around a specific chemical looping pilot plant constructed by the instructor and his team. Concepts of process safety analysis, safety relieve valve sizing, inherently safer design are directly exemplified in the pilot plant design making a clear translating class room lessons to real-world applications. Though this approach engages students by rooting the class on a specific and detail process, such approach requires additional challenges it requires the students to familiarize themselves with the process concept, learn how to interpret control and mechanical specification documents, and decipher piping & instrumentation diagrams. However, with this approach, students can be better prepared to enter industry. The learnings from the first course are reinforced in the second, where the students apply PHA and risk assessment techniques to their open-ended projects. The present paper will initially discuss the experience in developing material that incorporated the chemical looping pilot plant development in the process design teaching material. Experience gained from the last two years of instructor this course in regards to student response and course refinement are discussed. Finally, examples and perspectives of the applications PHA and risk assessments from the first course used in practic by students in their second course will be presented. An overall perspective on specific changes that will be implemented in Fall 2018 and direction of further safety material will be shared.

534917 Integrating Chemical Process Safety across the Curriculum

Abstract:

Process safety education and training at US colleges and universities is currently insufficient to meet industry needs.[1] Failure to prepare students with essential process safety training can and does lead to catastrophic and financially devastating events. In addition, failing to include mandated process safety in the curriculum can cause a program to lose ABET accreditation. Loss of accreditation means loss of competitiveness and will result in fewer enrollees, which could add up to millions of dollars lost in annual tuition per year. The goal of this project is to integrate process safety across the Northeastern University (NEU) Chemical Engineering (ChmE) curriculum to address these needs and minimize these risks.

Industry has identified eight necessary chemical process safety learning outcomes as specified by the Safety and Chemical Engineering Education (SAChE) Committee[2]: These outcomes span from an understanding of past events, to the hazards and impacts of chemicals and releases, and move towards understanding how to assess the hazards and risks associated with a process and to design or mitigate against them.

Through this project, the NEU ChmE curriculum was updated to include these eight outcomes, integrated across eight core ChmE courses, taught by 14 instructors, in order to evaluate the impact of the curriculum change. The student perceptions of their respective skills were measured before and after the redesign and their scores on assignments/quizzes/tests that related to process safety needs were also measured. In addition, this project produced the framework needed for future collaborative initiatives in chemical engineering education. Faculty annual reviews now include specific and measurable educational goals. A review system for faculty to measure their education outcomes and reflect on, and propose future changes has been implemented, and an annual faculty education retreat has been established to develop department wide educational goals. In addition, students have created an electronic portfolio to showcase the evidence of their effective progress.

1. Luo, Y., Westmoreland, P. R., Alkaya, D.,Alves da Cruz, R. V., Grossmann, I. E., Provine, W. D., Silverstein, D. L., Steininger II, R. J., Talbot, J. B., Varma, A., McCreight, T., Chin, K, Schuster, D. *Academia-Industry Alignment: Expectations about New Graduates, an NSF-Sponsored Study led by the American Institute of Chemical Engineers*; 2015.
2. Safety And Chemical Engineering Education Committee, (SAChE), [cited 2017; Available from: http://www.sache.org/SACHEGuidelinesForABET.pdf].

532442 Where Does Process Safety Fit into the Chemical Engineering Curriculum?

Abstract:

This talk will discuss some of the options that all chemical engineering programs should consider. Should process safety be integrated through the curriculum or mostly be focused into a single separate course? Should the greatest emphasis be placed on regulations, computations, or case studies? How does process safety link to other hard to place topics such as technical communications and professional ethics?

543062 Rapid Process Development - Bridging the Gap from Early Formulation Design to Integrated Continuous Drug Product Manufacture for a Dry Granulation Process

Abstract:

The adoption of integrated continuous manufacturing platforms has been pioneered in both academic and industrial environments in recent years.[1] The primary drivers for the adaption of integrated continuous drug product manufacture include: reduced processing time, reduced capital and operating costs, minimized scale-up risks, more consistent process control achieved via automation and opportunities for rapid process development.[2,3] Where the drug product is an oral solid dosage form, continuous tablet manufacture is made possible by a defined sequence unit operations that operate continuously â€" e.g. continuous powder feeding, continuous powder blending, continuous granulation, milling, tablet compression and coating. Ensuring the integrated operation of such unit operations is critical to the successful implementation of a continuously manufactured drug product.

Early formulation design and development is often performed on a small scale where discrete unit operations that are either performed in batch mode or inherently continuous in their operation (e.g. tablet compression, roller compaction, etc.) are used to produce the product. Significant effort is required to translate these non-integrated unit operations to integrated continuous unit operations working in concert in order to produce the final product for commercial manufacture.

In this work, we present the necessary workflow highlighting considerations and efforts needed to transfer the knowledge from early formulation design through to a fully integrated continuous dry granulation process, specifically, excipient and API powder through to coated tablets. The continuous dry granulation process is described. The methods used for non-integrated testing of unit operations are outlined. The connection between non-integrated testing and measurement and the selection of appropriate process parameters is discussed. Product control strategy and residence time distribution measurements are highlighted. Early considerations for the generation of a desired manufacturing range in line with critical quality attributes are also discussed. A process workflow is generated that provides a means for rapid process development from early formulation design to Integrated Continuous Drug Product Manufacture for a Dry Granulation Process.

References

1-Moghtadernejad, S. *et al.* A Training on: Continuous Manufacturing (Direct Compaction) of Solid Dose Pharmaceutical Products. *Journal of Pharmaceutical Innovation* **13**, 155-187, doi:10.1007/s12247-018-9313-5 (2018).

2-TeÅ¼yk, M., Milanowski, B., Ernst, A. & Lulek, J. Recent progress in continuous and semi-continuous processing

of solid oral dosage forms: a review. *Drug Development and Industrial Pharmacy* **42**, 1195-1214, doi:10.3109/03639045.2015.1122607 (2016).

3-Byrn, S. *et al.* Achieving Continuous Manufacturing for Final Dosage Formation: Challenges and How to Meet Them May 20â€"21 2014 Continuous Manufacturing Symposium. *J. Pharm. Sci.* **104**, 792-802, doi:http://dx.doi.org/10.1002/jps.24247 (2015).

525421 Developing a Loss-in-Weight Feeder Design Space Based on Performance and Material Properties

Abstract:

Abstract

Loss-in-weight feeders play a vital role in controlling content uniformity in continuous manufacturing. If, in a brief period of time, the feed rate of one ingredient changes with respect to the others, the change in concentration of the process stream can propagate throughout the system. Understanding how the feeder and the material behave is vital. Powder properties, such as density and cohesion, can cause large variability in the flow rate of ingredients fed from powder feeders. Also, one needs to understand the operational ranges of the feeder, i.e. percent motor power, which coincides directly with screw speed. Knowing an ideal range of operation and correlating powder properties to process performance can lead to determining the design space of the feeder accurately, and to faster optimization of feeder performance, which also saves money and material. In this work, a K-Tron KT-20 pharmaceutical loss-in-weight feeder has been evaluated with four common screw types (coarse auger, fine auger, coarse concave, fine concave), for various pharmaceutical (Avicel, Lactose Monohydrate, Magnesium Stearate, Acetaminophen, Prosolv, etc.) and catalytic powders, at multiple feed rates.

529956 Distillation in the Pharmaceutical Industry

Abstract:

Distillation in the pharmaceutical industry

Distillation is the process of separating the components or substances from a liquid mixture by selective boiling and condensation to increase the concentration of selected components of the mixture. The process exploits differences in the volatility of the mixture's components and this physical separation makes distillation a unit operation of practically universal importance.

Distillation plays an important role in the pharmaceutical industry where single stage batch distillations are frequently used and not the multistage distillation in columns studied at university. Distillation can be operated in batch or continuous modes, in single or multiple steps (put and take), and can be run for a variety of objectives: separation, solvent swap, drying from water, purification from volatile or non-volatile impurities and crystallization.

Distillation within the pharma industry is often used to solvent swap because the reaction solvent is usually different to the crystallization solvent or to concentrate a solution before crystallization. They offer many challenges, the environment of the Active Pharmaceutical Ingredient (API) and intermediates need to be controlled and the solubility limits known, impurity formation needs to be restricted by controlling the temperature and distillation time. Also, it is important to consider that dissolved species can affect phase equilibria.

Distillation can achieve different results depending on the conditions and equipment selected. There are a range of equipment and operating choices used at production and laboratory scale but often these are limited by the constraints from the chemistry involved. Phase equilibria is independent of scale, but other factors such as heat losses, vacuum or minimum and maximum stir volume are not. Distillation times can be increased considerably on plant, therefore it is difficult to scale up using experimental information. Phase diagram understanding is fundamental for distillation and simulations help in the decision-making process for optimal conditions and process specific requirements.

As an API moves through development, the scale and site of manufacture is subject to change. Understanding distillation and de-risking this operation by analysing the impact of different parameters such as pressure and equipment capability is fundamental in the technical transfer process. An overview on this fundamental unit operation focussing on different case studies showing continuous distillation, crystallization distillation and process optimisation through simulations will be given showing distillation lessons learn from the past and learnings for the future.

One case study will show a solvent swap simulated with Aspen Properties & Batch Modeller and demonstrate how modelling led to an improved process reducing both cost and environmental burden

Another case study will show how a put and take distillation which led to product oiling / gumming, was changed to a constant volume operation significantly improving robustness and product quality.

532501 Maintenance Management for the Sensor Network in Continuous Pharmaceutical Systems

Abstract:

The progress in the adoption of monitoring and control systems has accelerated the progress towards smart manufacturing in the pharmaceutical industry. Smart manufacturing necessitates reliable and accurate measurements from field devices combined with good quality estimates of unmeasured process variables. Thus, the sensor network which consist of equipment measurements, field devices, PAT tools, process models and data analytics is a fundamental requirement for robust process operations.

Data reconciliation exploits the redundancy in the sensor network together with process models to ensure the validity and accuracy of process measurements. The effective implementation of data reconciliation requires reliability in the raw measurements from PAT tools. However, since secondary pharmaceutical manufacturing primarily involves processing particulate materials, fouling of the inline and online PAT tools is inevitable. Corrective maintenance of fouled sensors can cause operational losses and uncertainties in manufacturing systems. Moreover, regulatory filings would require sensor network strategies for handling sensor faults where multiple sensors could fail simultaneously. For a smart pharmaceutical facility, it is imperative to anticipate and correct problems with the sensor network instead of only reacting to problems.

In this paper we investigate the impact of predictive maintenance of the sensor network in continuous tableting processes. The maintenance policy seeks to maintain measurement bias within acceptable limits such as to satisfy the conditions required for data reconciliation, using mechanistic models and material balances. The integrated framework schedules the anticipated downtime in maintaining every sensor â€" disconnecting the sensor or a group of sensors, cleaning the sensors and connecting it back to the sensor network. It is important to note here that the maintenance policy depends on the physical setup, materials used, the sensor network of the process and must rely on and learn from process history.

The challenges of implementing a robust and reliable sensor network are discussed, followed by demonstration of the effectiveness of the maintenance framework for a direct compaction case study. By integrating predictive maintenance with data reconciliation, the framework enhances the reliability in the measured estimates of CQAs for practical realization of Real Time Release Testing.

References

1. Smart Process Plants, M. Bagajewicz (2010)

533431 Leveraging Mechanistic Models for Scale up and Optimization of Lyophilization

Abstract:

Leveraging mechanistic models for scale up and optimization of Lyophilization.

Abstract

Generic pharmaceutical industry has to operate in tight timelines of product development and scale up from lab to plant where time to market makes or breaks the viability of the business case. Getting it right the first time demands detailed product & process understanding and implementing appropriate mechanistic modelling and Quality-by-design (QbD) approaches in the product life cycle, to minimize downstream risks. This work describes example cases of such efforts involving the freeze/ vacuum drying unit operation with solvents.

Lyophilization or freeze drying with aqueous/ solvent systems is a routine, but expensive, unit operation carried out across a range of industries (pharmaceuticals, food, paper, etc). It's especially relevant for temperature sensitive materials (for ex., peptides, liposomes, and onco molecules) where traditional high temperature drying unit operations cannot be deployed. The process involves loading the solutions of the required products in vials or trays and drying to obtain typically a free flowing powder (without meltback or collapse or oiling out). Typical challenges in this process involve: defining the optimal drying conditions (temperature, pressure vs. time) for a given material on a certain equipment so that product of an acceptable quality is obtained; ensuring uniformity of this quality across the vials in a given batch; and translation of such conditions during scale-up from lab scale to plant scale. For both the aqueous and solvent based systems, this demands a quantitative understanding of the impact of the complex interactions (Figure 1) of the relevant material properties, equipment characteristics, process conditions on the quality of the output

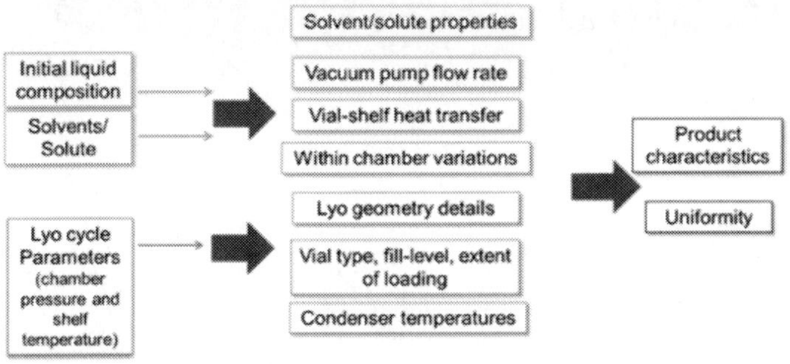

Figure 1 – Lyophilization- CMAs, CQAs, and CPPs.

product. This is especially true as the cost of failure at plant scale is prohibitive as this is typically the last step in the manufacturing of these products. Substantial literature exists (for example, M.J. Pikal, et al 1984, Pisano et al 2013) in capturing the vial-level physics with quasi-steady state assumptions to describe aqueous systems and demonstrating the applicability. These are solved using one-dimensional coupled heat and mass transfer equations.

The present work focuses on leveraging vial and equipment level physics to help with the definition and scale-up of lyo cycles in aqueous and solvent based systems. The first part focuses via case studies on how thermal characterization of lyophilizers can be leveraged to: bring predictability in scale-up (from 100s of vials to 1000s of vials) and; optimize lyo cycles using a novel algorithm. Example scenarios show such optimization lowering the drying time by as much as 20% compared to the traditional cycle development strategy. The second part presents cases of how such efforts can be extended to solvent systems by incorporation of additional physics. The higher vapor pressure of the solvent based systems demands more attention to the downstream utilities such as condenser and vacuum pump during scale up.

522002 Incorporation of Process Intensification Options in a Fed-Batch Cell Culture Manufacturing Platform

Abstract:

Several cell culture process intensification approaches were explored at AbbVie for increasing the productivity and processing speed of our protein biologics, with minimal impact on product quality profiles. Included in this are: the use of 10 mL cryovials to shorten the duration of seed train expansion, the use of perfusion technology to facilitate high cell densities at the seed bioreactor stage, and the use of perfusion and alternating tangential flow technology to facilitate increased protein yields at the production bioreactor stage. These approaches were demonstrated individually across all cell culture unit operations at laboratory and pilot scales. The work detailed in this poster provides a snapshot of the future of our biologics manufacturing capability as we continue to incorporate â€œpieces of the puzzleâ€ of process intensification for our biologics pipeline.

544816 CESMII Overview

Abstract:

The Clean Energy Smart Manufacturing Innovation Institute (CESMII), established by the US Department of Energy, is part of a network of institutes under the Manufacturing USA Initiative with a goal to secure the future of manufacturing in the U.S. through innovation, collaboration and education. CESMIIâ€™s mission is to radically democratize and accelerate the development and adoption of Smart Manufacturing (SM) technologies (sensing, control, data analytics, predictive modeling and platforms) to become the driving, sustainable engine that delivers energy efficiency and real-time business improvements in U.S. manufacturing. CESMII integrated strategies to achieve this mission include business practice innovation, enabling technologies, workforce development and a collaborative Smart Manufacturing Platformâ„¢ for industrial applications. Haresh Malkani, Chief Technology Officer â€“ CESMII, will provide an overview of the institute, its technology focus areas and recent progress.

544817 CESMII Roadmap Projects – Advancing CESMII's Mission

Abstract:

CESMII is about to kick off a portfolio of 10 innovative R&D roadmap projects covering a range of technologies including sensing, process control, data analytics, modeling, as well as workforce development, to address significant challenges faced by the manufacturing community. The projects cut across manufacturing types (discrete, hybrid, process), industry verticals (steel, cement, chemical, aerospace), and scale of focus (unit process, plant, enterprise). Miguel Corcio, Director - Technology Programs - CESMII, we will provide brief overview of the portfolio during this session.

546992 Panel Session: Democratization of Smart Manufacturing (SM)

Abstract:

Industrie 4.0, Smart Manufacturing, Industrial Internet of Things, Big Data, Cloud Computing are no longer just buzz words. They have become a reality and every year more manufacturers are getting started on this journey. But it takes investment in technology and talent to do so. Does it mean that this reality is only within the reach of the large and mighty companies? What can small and medium manufacturers do to be in this game? How do we make technology and expertise accessible for them cost effectively? Join our panelists representing both industry and academia in this healthy dialogue.

Panelists:

- Larry Megan – Praxair
- Sergio Butkewitsch-Choze – Arconic
- Panagiotis Christofides – UCLA
- B. Wayne Bequette – Rensselaer Polytechnic Institute

Moderator:

- Jim Davis, Vice Provost, Information Technology and Chief Academic Technology Officer, UCLA

527019 Practical Methods for Process Scale-up

Abstract:

Process scale-up in batch and semi batch industry (specialty chemistry, fine chemicals, active pharmaceuticals ingredient and many others) is characterized by a high level of uncertain and unexpected results. It is common to have processes where the purity of the final material and the in- process control tests are not reaching the specifications coming from the R&D procedures. As a result of this inconvenient situation, process transfer activities become difficult. In the R&D stage, the focus is to understand the mechanism of the process. whereas in the process transfer stage, the main target is to provide the conditions for the process in order to achieve the same results in yield and quality at all steps of the process. On the other hand, at the process transfer stage, quality of the raw materials and procedure of the process is well known, and normally is not the place to change them. As a consequence, the only reason for different results should come from the size of the equipment and from the different geometries used for the process in the upscale. It is clear from this forced change in the geometry, that the conditions provided to the process will change because of the difference in the flow generated at different sizes or geometries. It is also clear, that the process is progressing into the flow. So, if we will be able to identify which mixing parameters are significant for the process we will be able to avoid undesirable results.

In this presentation, we will discuss how to generate a good set of calculations and experiments that will provide a deep understanding of the scale up parameters that affect the process by understanding of the influence of the fluid dynamics and the mixing of the process.

516373 Commercializing the Production of 1,3 Butadiene through the Production of 2,3 Butanediol Using Intrexonâ€™s Natural Gas-to-Liquids Platform

Abstract:

The low cost and abundant supply of natural gas make it Americaâ€™s greatest untapped energy resource. Intrexon has developed the first natural gas-to-liquids bioconversion platform powered by methantrophic fermentation, applying synthetic biology to program bacteria to produce higher value materials of interest. The development of an innovative genetic toolbox enables one to engineer the biology of the methanotroph to more efficiently convert natural gas into a wide range of valuable chemicals and fuels.

This toolbox has enabled Intrexon to use its methanotroph platform to convert natural gas into 2, 3 butanediol and deploy a catalytic process to dehydrate 2, 3 butanediol into 1, 3 butadiene, an important chemical for the production of synthetic rubber. Using a methanotrophic bioprocess to produce 2, 3 butanediol has several advantages beyond the low cost and minimal processing of feedstock. First, natural gas bioconversion offers scaling opportunities beyond traditional gas-to-liquid (GTL) facilities, as well as biomass feedstock facilities. Second, the theoretical yields of methanotroph/methane exceed that of yeast/sugar approaches by a wide margin. Lastly, biocatalytic processes that achieve high yields under milder processing conditions allow for reduced capital and operation expenditures, offering greater economic viability.

To effectively bring this technology to commercial scale, Intrexon has successfully transferred technology from lab scale to the pilot plant scale. By operating extended runs at the pilot scale, safety, operability and technology readiness is rapidly addressed allowing the completion of the FEL-2 engineering package. The rapid pace at which the pilot work occurred came as a result of working with multiple partners who had both the assets and domain knowledge to operate parts of this process at the pilot scale. In addition to the consideration given to optimizing the bacterial metabolic pathway, key issues in the fermenter design, downstream processing and overall plant economics were addressed to evaluate all aspects of bringing this platform from lab to commercialization.

530072 Comparison of Distillation Strategies, Optimization and Scale-up for an Industrial Process Based on Mechanistic Modelling

Abstract:

APIâ€™s production/chemistry processes commonly include distillation operations for several purposes and in different stages and steps of the process. Despite not being considered as a critical or problematic operation, their study and mechanistic modelling can bring real gains to the process, including optimization of process conditions for different scales to save time and solvent volumes and reducing the number of lab experiments required.

Unlike the empirical or statistical models, the mechanistic model of a distillation considers and relates the mathematical descriptions of the chemical and physical processes. Thus, the model provides accurate answers to important questions raised when developing and scaling-up a distillation and allows a deeper knowledge of the process.

To increase the confidence level of the results provided by the mechanistic approach, we demonstrate our methodology in terms of equipment characterization, fluid properties calculation along the distillation and simulation of different distillation strategies (e.g. continuous feed distillation vs put-and-take) while developing our models using DynoChem. Distillation data gathered from previous experimental runs was used to validate the mechanistic model by direct comparison of key process variables.

We demonstrate the application of this methodology by showing case studies with different objectives for industrial projects. In one of the examples, we present the comparison between different distillation strategies for specific solvent swap in the process where we pretend to achieve a target final composition while reducing the volumes of fresh solvent needed without impact the process outcome. A second application is focused on the scale-up from Laboratorial to Pilot Scale and optimizing the distillation conditions to reduce operation time through simulations.

532930 Development of Mobile Decontamination Equipment for Hazardous Materials

Abstract:

A rapid emergency response is the most important task to prevent toxic dispersion when hazardous materials are released. This study focused on the development of mobile decontamination equipment which can respond immediately to remove or neutralize released hazardous materials. The developed equipment is novel in that it combines collecting the hazardous materials and dispersing neutralization agents in one system. In 2013, the Korean Ministry of Environment designated 42 hazardous materials, mostly in liquid phase, based on the risk to the environment, the risk to human, the frequency of accidents, and the volumes handled. Accordingly, Korean government established a 6 million US dollar project for developing mobile equipment for the decontamination of hazardous materials. Various prevention techniques and existing mobile decontamination equipment were studied to help develop efficient equipment. After considering the corrosion and fouling effects of designated hazardous materials, carbon steel was selected as the material of construction with Teflon coating to exposed applied portions. The equipment, designed to handle up to 30 m^3 of hazardous materials, was comprised of four truck mounted units; spill collecting, neutralization tank, dispersal, and neutralization agent tank. The operation was as follows. First, the spill collecting unit collects a large amount of released liquid hazardous material and it is neutralized in the second unit. After collecting, the dispersal unit disperses neutralization agent to neutralize remaining hazardous materials in field. In this study, pilot scale equipment was developed that its dispersing distance and the rate of collecting was tested. In addition, neutralization agents were sprayed to the various acids for the test which was conducted in 7 m^3 size facility. Test drew several important points that are effective for improving quality of the final product. In conclusion, this study could help reduce toxic dispersion which deals with hazardous material releases.

Process Optimization to reduce urea plant startup failure rate and ammonia pollution

NAVEED RAZA 1, ALI AYUB2, KASHIF JAMEEL3

BILAL MUSTAFA4,
ENGRO FERTILIZERS LIMITED
DAHARKI, PAKISTAN

Abstract

This document explicates the improvisations made to overcome the startup failures due to hardware limitations of Urea Plant at Engro Fertilizers Limited, Pakistan. Since plant commissioning, the success rate of the Urea plant startup has been low because of instability of the high pressure scrubber U-seal which also results in ammonia pollution & excessive venting. Startups experienced during the operating period of more than 20 years have been listed. Each scenario was studied with thorough Root Cause Analysis, resulting in identification and implementation of process variations. These changes resulted in improved startup success rate and reduced startup time. This eventually helped in minimizing the excessive venting and therefore eliminated environmental concerns which ultimately facilitated Engro to keep their promise of Clean & Pollution Free Environment.

Keywords: Urea, Startup, TOYO ACES, Environment Pollution, Stripper.

1. INTRODUCTION

Engro Fertilizers Limited, a wholly owned Engro Corporation subsidiary, is a premier fertilizer manufacturing industry in Pakistan. In 1992, Engro relocated a 1,000 t/d urea plant, from ICI Billingham, UK, to its manufacturing site in Daharki, Pakistan. The Toyo Total Recycle C urea plant from ICI was commissioned in 1993. Ever since its commissioning, the urea plant has undergone various debottlenecks and capacity enhancement programmes, energy index improvement schemes and various environmental control measures. The relocated urea plant now employs Toyo's ACES process to manufacture prilled urea. Since the commissioning of the TOYO[1] ACES[2] plant in 1998, the success rate of the Urea plant.

Startup has been low, due to instability of the high pressure scrubber U-seal. With the passage of time the issue became more agitated due to Carbamate Condensers tubes plugging and stripper inefficiency. The major causes of scrubber U-seal failure during startup is reactor overflow to scrubber due to stripper pressurization. With the failure of the High pressure scrubber U-seal, feeds need to be cut, followed by reactor draining, and flushing before the feeds are fed again. This produces a lag of 8-12 hours once Startup fails hence inculcating in production loss. Listed in the paper are the Improvements which were made in the standard Startup procedures, and operational techniques.

1 TOYO is an Engineering and construction Corporation for Industrial Facilities

2 ACES (Advanced Process for cost and energy saving) is a Urea manufacturing technology by TOYO Engineering Corporation.

2. BRIEF DESCRIPTION OF ACES PROCESS

The TOYO ACES Process is to minimize energy input to the urea plant by combining the features of solution recycle process and stripping process i.e. a high CO_2 conversion and the highly efficient separation of unreacted materials. The ACES Process drastically reduced steam consumption compared to Total Recycle Process. The Urea-2 plant ACES loop is designed to operate at isobar equivalent to 174 kg/cm^2. The plant is operated at N/C [4] = 4.0, H/C [5] = 0.79 and 66% CO2 conversion is attained at the reactor outlet. The high CO_2 conversion reduces the energy required for decomposition of unconverted materials. The proprietarily designed tray-falling film Stripper efficiently decomposes and separates ammonium carbamate and excess ammonia in urea synthesis solution from the reactor. In Carbamate Condensers, the stripper off gasses are condensed and absorbed by the recycle solution from the recovery section. The heat of condensation and absorption is utilized by generating 5.0 Kg/cm^2 steam in No.1 condeser and for heating the stripper outlet urea solution in No.2 condenser. Both gas and solution at the bottom of the carbamate condensers is fed to the reactor. The reactor top gas, containing small quantities of ammonia and CO_2 is absorbed in HP scrubber using the recycle solution, which then headways to reactor. Fig. 1 shows the simplified flow diagram of the ACES Process synthesis section.

Fig 01: Urea Plant Process Flow Diagram

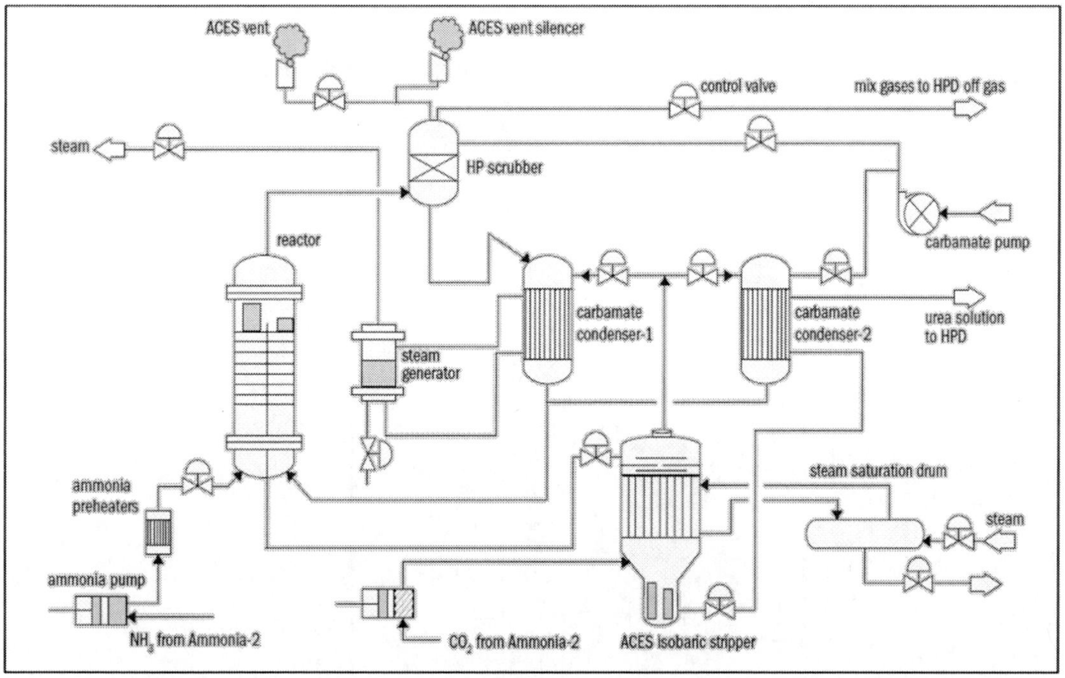

4 N/C (Ammonia to Carbon dioxide ratio) is one of the controlling perimeter of equilibrium reaction.

5 H/C (water to Carbon dioxide ratio) is one of the controlling perimeter of equilibrium reaction

3. STARTUP FAILURE ROOT CAUSE ANALYSIS AND IMPROVISATIONS:

High pressure scrubber U-seal instability has always remained a challenge for Urea plant technical staff. Multiple brainstorming sessions and root cause analysis were carried out to identify the Key factor for Startup failure. Today our efforts have led us to overcome this mystery. Improvements have been made in the standard operating procedures and operational techniques to improve the success rate of urea plant Startup based upon experimental results. Below is a complete description of the research that has been carried out which have actually decreased Startup failure frequency.

3.1 BYPASSING STRIPPER DUE TO CARBAMATE CONDENSER TUBE PLUGGING:

Once the reactor overflows during ACES Startup the reactor solution is transferred to stripper via reactor U-seal. Reactor solution contains urea, unconverted ammonia, carbon dioxide and water. These uncovered gases are stripped off in the stripper and shifted to carbamate condensers. Carbamate condensers control the ACES loop pressure by condensing the stripper off gasses. Due to tubes plugging of Carbamate Condenser-1 & Carbamate Condenser-2, their heat duty has decreased significantly. This leads to stripper pressurization as the gasses are not condensed in carbamate condensers. Pressurized stripper produces a high back pressure on reactor U-seal which doesn't allow reactor solution to flow to the stripper hence leading to reactor level hold up. A high Reactor level increases ACES loop pressure, while further increase in level lets the reactor solution to overflow to scrubber, resulting in ACES Startup failure due to ACES loop high pressure or scrubber U-seal failure.

Fig 02: Bypassing stripper by using reactor drain valve

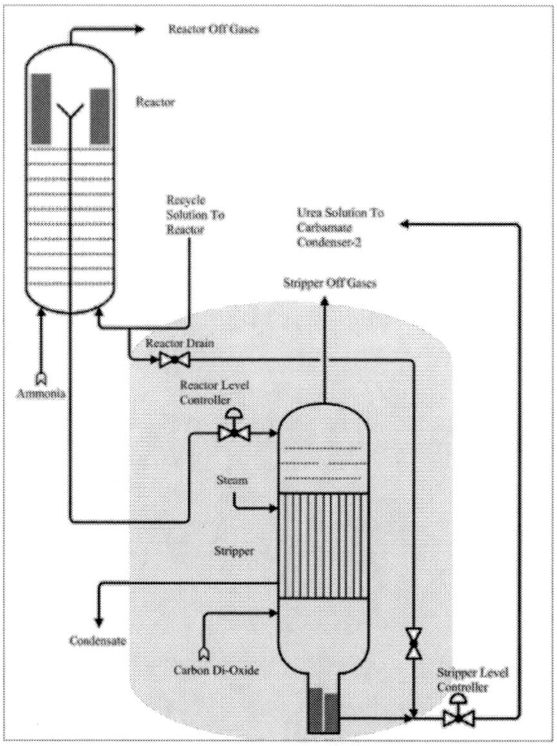

3.1.1 Remedial Action:

Stripper over pressurization is the key factor contributing towards reactor level hold up. An improvised way to maintain reactor level is bypassing the stripper through reactor drain valve during plant Startup as shown in figure 02. During Startup flow rate of the stripper off gasses is high due to low first pass conversion through the reactor as the reaction is not well established. By operating the reactor drain valve, lesser solution will flow to the stripper hence it will not be pressurized, at the same time this will help to control reactor level. After this improvisation, Urea plant Startup failure rate has decreased significantly.

3.2 Steam generation drum pressure controller set point reduction:

Carbamate Condenser No. 1 (CC-1) is used for condensing stripper off gasses in the tube side. The heat generated by the formation of carbamate solution in the tubes is utilized in generating 4.92 kg/cm2 steam in the shell side. This is achieved by circulating steam condensate from steam generation drum through the shell side of carbamate condenser (CC-1). A pressure controller is installed at the steam outlet line of steam generation drum to control the pressure of drum. Varying the set point of steam generation drum pressure controller has a direct impact on condensation of the stripper off gasses inside the Carbamate condenser-01 tube side. The gasses generated in the stripper should be condensed inside the Carbamate condenser-01 and Carbamate Condenser 02 else they will produce a back pressure on the scrubber U-seal leading to ACES Startup failure. We are facing similar problem these days due to reduced heat duty of Carbamate condenser-01 and Carbamate

Fig 03: Steam generation drum pressure controller set point reduction

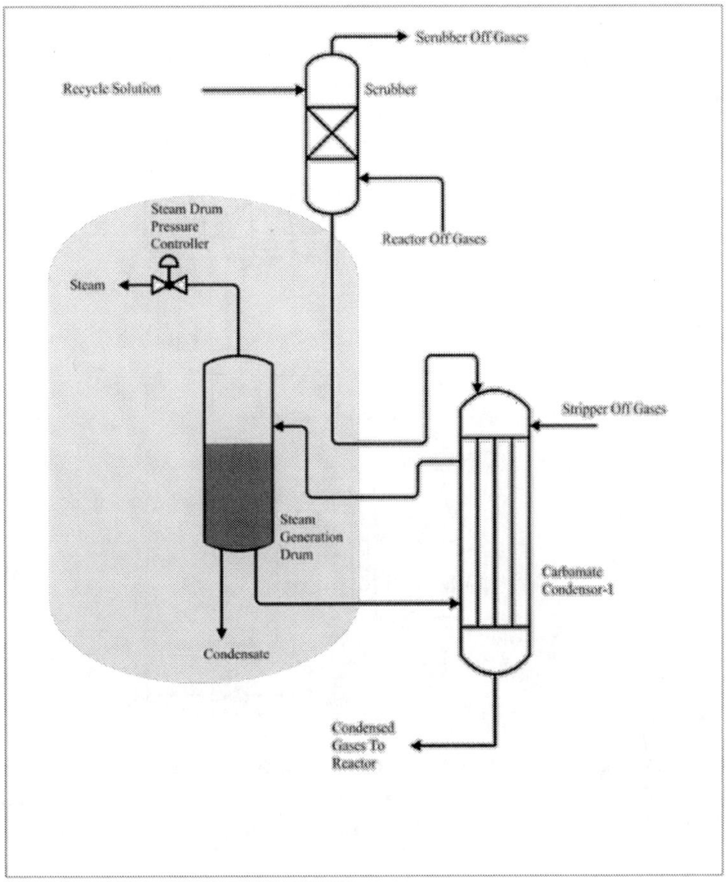

Condenser- 02 because of their tubes plugging. Figure 03 describes the pictorial view of the above mentioned process.

3.2.1 Remedial Action:

As per normal Startup procedure the pressure of steam generation drum should be 5.27 kg/cm2 while during normal plant operation it is controlled at 5.97 kg/cm2. To overcome the scrubber U-seal failure due to CC-1 tubes plugging and high vapour load during Startup, set point of steam generation drum pressure controller is reduced from 5.27 kg/cm2 to 4.56 kg/cm2. Decreasing the set point increases the heat transfer across Carbamate condenser-01. The impact of this improvisation has been successfully realized and is now the part of our standard Startup procedure.

3.3 Scrubber U-seal temperature profile variation and control:

Scrubber U-seal is in place to provide the necessary head pressure required for the carbamate solution to restrict stripper off gasses and flow inside the Carbamate condenser-01. During Startup when reactor overflows into the stripper it generates lots of vapours due to very low first pass conversion through the reactor. Under high vapour load it becomes difficult to maintain the scrubber U-seal flow, as vapours back pressure overcome the head pressure of the U-seal. The success rate of the Startup is highly dependent on the scrubber U-seal stability. Instability causes high pressure in the ACES loop leading to Startup failure.

Fig 04: Scrubber U-Seal temperature profile variation and control

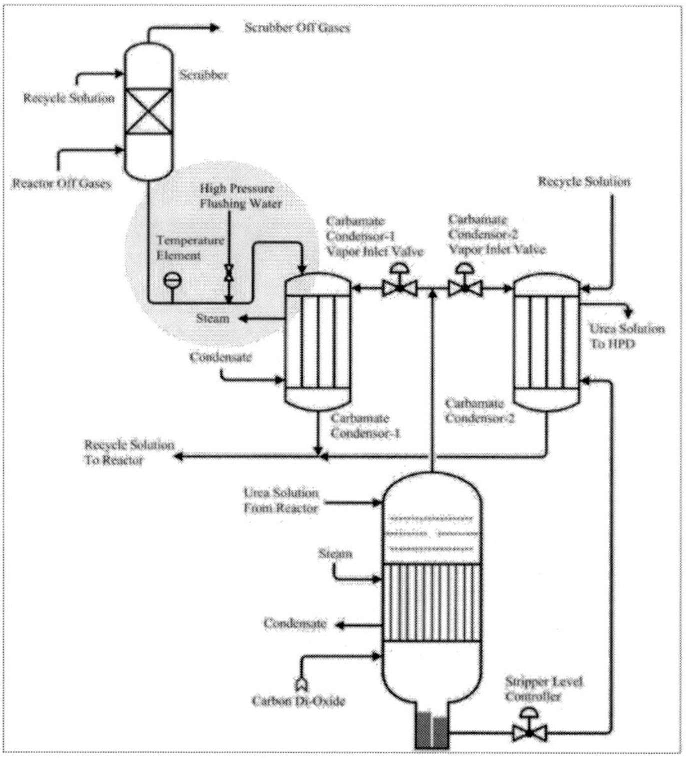

The probability of scrubber U-seal failure is greater in Startup as the reaction is not well established and quantity of unabsorbed gasses inside the loop is greater. For that reason during Startup a continuous supply of HP Flushing water is lined up in the scrubber U-seal, as rate of absorption of

unabsorbed gasses will be greater in a lean solution. If in any case the gasses inside the Carbamate Condenser are not condensed, they will produce back pressure upon the seal.

This leads to restriction of flow inside the scrubber U-seal. If the liquid holds up, the solution required for gasses condensation inside the Carbamate condenser-01 will not be available hence an abrupt increase in loop pressure will occur leading to scrubber U-seal failure. Variation in the Temperature element installed at Scrubber U-seal explains the clear picture of level hold up in U-seal. The phenomenon of variation in temperature due to flow restriction in Scrubber U-seal is known as kink and has become more prominent these days due to reduced heat duty of Carbamate condenser-01 because of its tubes plugging.

3.3.1 Remedial Action:

As per normal plant Startup procedure stripper off gasses inlet valve opening towards carbamate condenser-01 is 90 percentage and this opening remains constant. During Startup, a downward kink in U-seal temperature is observed due to restriction of recycle solution through the scrubber U-seal, because of high vapour load and low heat duty of carbamate condenser-01. To re-establish the recycle flow, stripper off gases inlet valve opening is decreased from 90 percentage to 80 percentage. This change will decrease the vapour flow through Carbamate Condenser-1. Consequently, recycle flow is recovered through the U-seal and later its opening is normalized. The impact of this variation has been successfully realized and is now the part of our standard Startup procedure. Scrubber U-seal temperature profile variation (in green colour) is shown in Fig 5 with respect to stripper off gasses inlet valve opening (in white colour). This trend is taken from the DCS snap shot of the recent Startup. Trend shows stability in Scrubber U-seal temperatures once the load on the Carbamate Condenser-01 is reduced by decreasing the stripper off-gasses flow via stripper off gasses inlet valve to CC-01.

Fig 05: Scrubber U-Seal temp variation V/S Gasses flow to Carbamate condenser

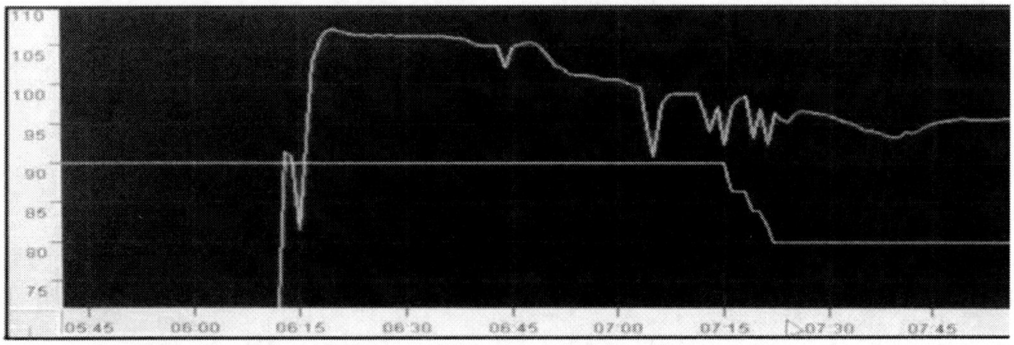

3.4 Startup with improved urea concentration feed solution:

Earlier practices were to start urea plant by feeding urea solution of lower concentration to the reactor i.e. 15-25%. A low concentration means greater quantity of water in the system. As the reaction conversion in already low during Startup, a higher H/C ratio will aggravate the situation. As the loop pressure is controlled by reaction conversion and condensation, a low reaction conversion will lead to higher vapour load and eventually lead to scrubber U-seal instability.

3.4.1 Remedial Action:

- Level make-up of low pressure and high pressure absorbers with 25-30% concentration by weight of urea solution helps to overcome the defined issue.
- Increased liquid density helps in maintaining the U-seal head pressure, as concentrated liquid mass flows in counter current pattern with the vapours flowing into Carbamate condenser-1.

4. AMMONIA POLLUTION CONTROL:

The urea plant has 83.8 meter high atmospheric vent to relieve pressure from the high pressure loop if required. Urea plant Startup is renowned for its excessive venting and failures due to scrubber U-seal failure resulting in carbamate showering from the ACES vents.

With improved environmental standards and stringent checks on venting, modern plants have developed administrative controls and engineering checks like atmospheric scrubbers or flare systems to avert the release of hazardous chemicals into the open air.

A low cost and effective solution was a big challenge for the urea plant team. Environmental complaints from the surrounding area and within plant premises became a routine during and after Startup due to the frequent Startup attempts and improved community awareness about environment checks.

As per standard operating procedure it is recommended to start the urea plant with 40% load and 15-25% carbamate solution concentration. Reactor feed rates, overflow time, and vapour dynamics inside the ACES loop are calculated on this basis. This happens because, unlike other plants, steam to the stripper cannot be increased immediately once the reactor overflows due to the panic of U-seal failure. The load is hence shifted to the downstream circuit which resulted in increased pressure and thus venting.

As the stripper liquid outlet valve is opened to release solution from 100 kg/cm^2 pressure to the HPD it was a challenge for plant operators to maintain the HP circuit pressure within limits i.e. 100 kg/cm^2. Due to the high feed rate of solution equivalent to 40% load coming from ACES and low conversion of initial pass solution, vapour load on the HP circuit is very high, which triggers the opening of the atmospheric vent to release ammonia and inerts to control HP loop pressure. Almost 30 Ton/hr of the vapours escape from the atmospheric vent, releasing ammonia with a concentration of 99% to the environment.

4.1 Remedial Action:

- Today urea plant is started with 30% load basis instead of 40% to avert the uncontrolled condition of rapid vapour generation once the stripper liquid outlet valve opens and the first pass of solution reaches the HPD.
- Early line-up of ammonia refluxes in the high pressure absorber helps to minimize the opening of the atmospheric vent to release inert gasses that are initially used to pressurize the HP circuit as a Startup permissive.

With the control on the HP circuit pressure, the reflux ammonia condensers and absorbers are lined up during Startup. This helps to recover ammonia instead of venting to atmosphere. So that environmentally hazardous vapours are recovered and returned back to the process. This reduces the number of environmental complaints from the local vicinity and other operating areas to zero from five as recorded in the previous year.

5. CONCLUSION:

A remarkable increase in Urea plant Startup successful rate has been achieved through various improvisations and variations in the standard Startup procedure and operational techniques. Even though Low heat duty of Carbamate condenser and inefficiency of stripper opposed much to the cause, still we managed improving the success rate of Urea Plant Startup, reduction in Startup time and quick Startup post seal failure scenario. At the same time reduced excessive venting and hence environmental concerns.

520386 Launching Clinical Antisense Oligonucleotide Manufacturing Capability in a Biologics Company

Abstract:

In 2013, Biogen entered into a series of agreements with Ionis (then Isis) Pharmaceuticals to develop a pipeline of oligonucleotide-based molecules to treat neuroscience diseases like Spinal Muscular Atrophy, ALS, and Alzheimerâ €™s. Since the launching of this partnership, Biogen and Ionis have successfully brought to market a groundbreaking therapy for SMA and have transitioned six other oligonucleotides from research to development. In 2016, the antisense oligonucleotide manufacturing industry was saturated with demand as this industry began to take off. Based on limited existing capacity in ASO manufacturing and Biogenâ€™s growing ASO pipeline, the company made the decision to internalize the capability to supply ASOs for their clinical and commercial products. Biogen is the first innovator to invest internal capital on an ASO facility, while others have chosen to outsource. By vertically integrating their ASO supply chain, Biogen has been able to significantly reduce turnaround time and reduce supply risk.

Biogen has historically been a protein therapeutic company, however in the last several years they have broadened their scope to include small molecules, antisense oligonucleotides, and gene therapies. However, their business processes were built to support biologics. There were many challenges associated with internalizing a new manufacturing platform inside a biologics company, which required creative solutions by the project team and focused execution. Among those challenges was establishing a manufacturing operations staff using only internal staff who had experience in biologics or small molecules, but not in ASOs. Staff selection, structure and culture building were key in enabling team success. The team leveraged their partnership with Ionis and with their technical development colleagues to accelerate learning concurrently with starting up a new factory.

The startup activities occurred over a period of about 17 months from capital funding approval to completion of the first clinical batch. The construction and tech transfer teams were closely integrated throughout the entire project to ensure a coordinated approach and to meet the extremely aggressive project schedule. The first batch demonstrated Biogenâ€™s capability of manufacturing quality antisense oligonucleotides in their new manufacturing facility.

521955 Reaction Engineering for Energetic Material Synthesis

Abstract:

Conventional manufacturing of energetic materials is dominated by the use of batch and semi-batch reactors. These methods offer flexibility at the cost of safety, reproducibility and potentially higher cost products. Continuous processing offers superior safety, ease of scale-up, lower cost and access to new materials. This presentation will discuss the use of process analytical technology (PAT) to interrogate reaction processes applied to energetic materials. The motivation is to assess their amenability toward continuous processing methods.

530209 Development and Evaluation of Oxidative Dehydrogenation Technology for Production of Light Olefins

Abstract:

Light oleï¬ns are traditionally produced by the steam pyrolysis of saturated hydrocarbons as well as by recovery from refinery cracking operations. These technologies results in a wide range of byproducts, some valuable and some that need disposal. All of the conventional technology require extensive separation and purification.The technology known as RMD for reaction, membranes and distillation utilizies the M1 catalyst developed by Idaho National Laboratories and features very high selectivity at various levels of conversion in studies on the conversion of ethane to ethylene. Additionally, INL has found that the produced ethylene is refractory in additional reactors. This allows the interchanging of reactors and membranes, both of which provide high concentrations of ethylene. Process development and the associated economic analyses were performed using a cost of production procedure (COP) commonly called a required netback (RNB) analysis. These results were analyzed by incorporating the expected performance of a variety of $R_xM_yD_z$ arrangements with x,y, and z representing the number of the process segments. So for instance, $R_2M_1D_1$ indicates 2 ODH reactors, 1 facilitated transport (FTS) membrane and one heat pump distillation section. Results indicate that the combination of ODH reactors with FTS membranes is a new route that has much lower variable costs, ï¬xed costs and capital recovery that can reach shutdown economics level versus some existing technology.

530251 Process Intensification through Membrane Separation of Olefins and Paraffins

Abstract:

Ethylene and propylene are two of the major building blocks for the chemical industry. The separation of ethylene and propylene from their alkane counterparts is highly energy-intensive. It is estimated that separating alkanes from alkenes through distillation uses 0.3% of energy globally. In addition, distillation columns to separate these molecules are very capital-intensive. A membrane based olefin-paraffin separation technology has the potential to provide substantial economic benefit to the petrochemical industry and to drastically reduce the energy required for this separation. In addition, membrane systems are scalable, allowing their implementation for smaller, stranded streams where distillation columns have not previously been installed due to limited economic value. Membrane processes utilizing facilitated transport membranes have been extensively studied for separating olefin/paraffin mixtures. While separations have been demonstrated in the laboratory, problems with membrane stability have prevented development of commercial systems.

Compact Membrane Systems (CMS) has developed the Optiperm™ membrane, a customized amorphous fluoropolymer (CAF) containing silver ions that selectively transport olefin molecules from a mixture of olefin and paraffin. The technology exploits a reversible complexation reaction mechanism known as facilitated transport. Silver incorporated into the membrane polymer backbone acts as a binding site for the double bond in the olefins transporting them across the membrane as depicted in the cartoon below. Saturated paraffins do not have the same interaction. This difference in affinity allows for olefins to pass through the membrane at a significantly higher rate than do paraffins.

This talk will share the latest field results from both the pilot currently running at the Delaware City Refining Company as well as any results coming out of the RAPID-funded collaboration with Dow Chemical. The DCRC pilot is operating on a stream with a 10-20% concentration of propylene while the Dow Chemical work will focus on higher olefins concentrations of both C2 and C3 streams. This will include performance data, longevity and membrane stability results as well as any learnings from these projects. Thus far, the DCRC pilot has demonstrated over 45 days of stable membrane operation with propylene permeance >75 GPU and selectivity >20 for propylene over propane.

535284 2D Optimization of Fixed-Bed Reactors: Additional Degrees of Freedom for the Reactor Design to Increase Efficiency

Abstract:

Highly efficient reactors are fundamental for resource- and energy efficient production of chemicals. To enable high reactor efficiencies, the reactor design aims to realize optimal reaction conditions along the reaction route [1]. In this regard, the reaction temperature plays a superior role due to its strong influence of the reaction rates.

For heterogeneously catalyzed reactions with a pronounced energy release, the state-of-the-art approach is the use of tube bundle reactors with axial dilution zones to approximate the optimal temperature [2]. However, a significant deviation from the optimal temperature can still exist in radial direction of the tubes due to the radial heat transport resistances as shown in figure 1.

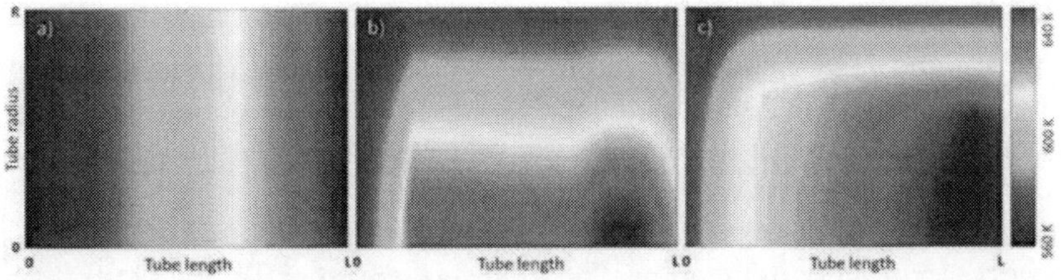

Figure 1: Exemplary demonstration of the (system specific) optimal reaction temperature along the tube (a) compared to the resulting reaction temperature profile due to heat transport resistances in a fixed-bed reactor with an optimal axial catalyst bed dilution profile (b) and a fixed-bed reactor with an optimal axial and radial catalyst bed dilution profile (c) for the air-based production of ethylene oxide.

This deviation can be reduced by further increasing the surface-to-volume ratio of the tubes, thereby reducing the radial transport pathways. To maintain the necessary residence time, this approach requires either longer tubes or an increase of the number of tubes. Both options lead to an increase of investment and operation costs as a result of a higher pressure drop and higher material and/or installation costs.

As an alternative, the concept of the radial optimization of tubular reactors is introduced for the model-based reactor design. This is an extension to the state-of-the art approach of optimizing suitable design variables such as the catalyst bed dilution or the catalyst pellet specifications in axial direction of the tube [3]. The radial coordinate is additionally considered as a degree of freedom for the chosen design variables, leading to a superior approximation of the optimal reaction temperature as shown in figure 1 c). The optimization results give a quantitative indication of the potential reactor performance gain by introducing radial profiles of the chosen design variables. It can then be investigated to which extent these profiles can be technically approximated. Furthermore, the optimization results can indicate the potential of suitable alternate reactor concepts (e.g. catalytic foam reactors or monolithic reactors).

The concept of the radial optimization is presented and illustrated by optimizing a tubular reactor for the air-based production of ethylene oxide in a case study. To reduce the computational effort for the resulting large-scale nonlinear optimization problem, pseudo-homogeneous reactor balance equations are used to describe the reactor. The solutions of the catalyst pellet equations are approximated via analytical approximations [4]. A suitable model basis is identified that is compatible with the dynamic optimization approach, yet considers changes of design variables in radial direction of the tube. The objective function is to maximize the conversion of ethylene while fulfilling other reactor performance constraints in terms of selectivity towards ethylene oxide and space-time yield. Results show that by introducing a 2D dilution profile, the yield of ethylene oxide can be increased by 2 %. Furthermore, the 2D optimization results indicate that a novel alternate reactor concept consisting of a catalytic foam reactor followed by a classical particle packed fixed-bed reactor holds significant potential for process intensification.

References

[1] Peschel, A.; Freund, H.; Sundmacher, K.; Ind. Eng. Chem. Res. 49 (2010) 10535-10548

[2] Eigenberger, G.; Ruppel, W.; Ullmann's Encycl. Ind. Chem. Online ed. Wiley-VCH, Weinheim, 2012

[3] Hwang, S.; Smith, R.; Chem. Eng. Sci. 59 (2004) 4229-4243

[4] Pietschak, A.; Kaiser, M.; Freund, H.; Chem. Eng. Res. Des. (2018) submitted

534776 On-Sun Demonstration of Continuous Redox Driven Solar Thermochemical Hydrogen Production

Abstract:

A sustainable hydrogen (H2) economy based on the low-cost production of H2 from renewable resources has the potential to drastically transform the energy sector. A simple two-step reduction/oxidation (redox) cycle allowing for the production of oxygen (O2) and H2 by solar thermal water splitting (STWS) has been demonstrated on-sun. The H2 and O2 are produced in two separate steps, eliminating the need for high-temperature H2/O2 gas separation. In this redox process, oxygen is generated by the high-temperature reduction of an active metal oxide spinel, forming O-vacancies; followed by a second step in which the vacancies are filled during a high-temperature steam exposure, producing H2 fuel. Although the STWS reaction scheme appears straightforward, its implementation as an economically viable process is extremely demanding. In order to commercially realize this technology, several challenges must be overcome including an energy efficient reactor design, stable reactor and heat exchanger materials, robust redox active particles, and an efficient process for recycling inert gas. Here, we present results showing near-isothermal on-sun H2 production utilizing dual lab-scale fluidized beds operated in which redox cycling produces H_2 continuously. Active hercynite particles were redox cycled at the High Flux Solar Furnace at the National Renewable Energy Laboratory. Production rates of H_2 exceeding DOE project targets and a final deliverable of H2 production were demonstrated. The impact of incident solar radiation and hydrogen production is evaluated during on-sun testing. Results of a techno-economic analysis will be discussed as will the path forward. This work demonstrates the viability of STWS leveraging continuous operation of both reduction and oxidation at near-isothermal conditions within a multi-tube solar thermal cavity reactor. This is the first demonstration of continuous STWS to be reported.

516776 The Mechanism Investigation for Process Intensification of the Esterification Reaction By Microwave Irradiation

Abstract:

Microwave has long been recognized to accelerate the rate of chemical reactions. However, the scientists still clash on the question what is main reason of the microwave intensification effect for this acceleration. Microwave changing the relative volatility of some special mixtures discovered by our previous research may open up new insight. For some reactions that have significantly limited by chemical equilibrium, if the product concentration in reaction system can be reduced through the changing the relative volatility of reaction system, the acceleration of reaction rate can be occurred by the special effect of microwave. In this paper, esterification, a typical equilibrium limitation reaction, has been chosen to determine the intensification effect of microwave. The experimental equipment installs a drainage system to move the water that is one of the reaction products. According to compare the experimental results under microwave and traditional heating, the hypothesis can be verified. Moreover, the main influence factor for reactive conversion, microwave power and reaction temperature were systematically investigated in detail to build the kinetic of this reaction in microwave. Experimental results show that microwave improved drainage can realize the reactive acceleration. The intensification effect of microwave for some reactions also can be explained by the investigation results.

532048 Process Intensification Via Batch-to-Continuous Transition in the Production of Lubricants: Acquisition of Robust Reaction Kinetics

Abstract:

Production of specialty chemicals to-date is conducted almost exclusively in batch or semi-batch reactors. Transition to continuous processing promises strong improvements—so-called "process intensification"—over the century-old batch processing technology with lower unit energy consumption, less waste generation, reduced exposure hazards, smaller process footprint, lower capital investment and more consistent product quality. The transition to continuous processing—already demonstrated widely at the much smaller scale of the pharmaceutical industry—thus offers the prospect of truly revolutionizing the manufacture of specialty chemicals.

As a first step, this transition requires the acquisition of robust information about the process chemistry. While batch reactors are highly tolerant to incomplete knowledge (in particular regarding kinetics), the design and operation of continuous processes requires availability of robust kinetics. However, the detailed chemistry and kinetics underlying current processes are in most cases poorly understood (if not completely unknown). Transition to continuous processing hence requires collection of robust and accurate kinetics as a key prerequisite, including impact of impurities and contaminants which play a significant role in many specialty chemicals processes.

This work presents results from a university-industry collaboration between Lubrizol Corporation and the University of Pittsburgh, under the umbrella of the recently formed RAPID Modular Chemical Process Intensification Institute. It aims to demonstrate the transition from batch to continuous processing focused on lubricant chemistry. Specifically, we present results from a kinetic investigation into the production of succinimide dispersants via reaction of a poly-isobutylene succinimide anhydride (PIBSA) with polyamines. The molecules undergo a condensation reaction to produce a large surfactant molecule with a polar head group and a long, oil soluble polymer tail. A complete family of products (which form the largest volume oil additives) is made by varying molecular weight, structure and stoichiometric ratio of the various reactants. However, the reaction kinetics for this class of reactions are unknown and the acquisition of robust kinetics is hence a crucial first step in the process development for a continuous process. We demonstrate that the combination of online IR spectroscopy as the analytical tool, with careful data evaluation (including model cross-validation techniques and analysis of transient traces) allows for the required acquisition of robust reaction kinetics and the identification of optimal operating windows for a continuous process.

542620 The Impact of Shale Gas and Oil on the Chemical Industry

Abstract:

Lower alkanes and aromatics are very useful as both fuels and chemical raw materials. Occasionally supply disruptions constrained availability but also encouraged investigation of alternative feedstocks and chemistries. Recent new technologies for the production of unconventional gas and oil sources from tight shale formations significantly increased availability of both methane and condensate, radically altered process economics, resulted in the construction of much new intermediates capacity, and multiplied the ways traditional and alternative chemistries might be exploited. This presentation will outline historically how natural gas components became useful as chemical feedstocks, how government regulations occasionally led to perturbations in availability and price, how supply crises led to the development of alternative chemistries, catalysts, and ultimately the exploitation of shale resources. More specifically, will be discussed how excess coproducts, especially condensate, from increasing shale gas and oil production for energy are having significant unexpected impacts on the chemical process industries and opportunities for new process chemistry research.

541186 Sustainable Energy and Chemicals: Past, Present, and Future

Abstract:

The last 40 years has seen the U.S. energy situation vacillate from gasoline rationing and concerns over Peak Oil, to projected energy and feedstock prices that were noncompetitive in global markets for chemicals production, to the economic revolution of frack gas and light tight oil. The importance of low cost energy and feedstocks in driving economic opportunity will be discussed relative to this history, along with the importance of sustainability in decision making to gain stakeholder buy-in and acceptance. Strategies for managing career paths and competencies despite changing targets and high uncertainty will be addressed.

542618 Disruptions: What the Future May Hold

Abstract:

Disruptive forces are all around us. They change the way we communicate and work. They change the way we develop goods and services. They also change the types of products that are manufactured. Examples such as the way in which we produce and consume energy or the feedstocks we use to produce product have seen many changes over the course of the last few decades, and these disruptions are expected to continue. The chemical engineering discipline and the expectations for a chemical engineerâ€™s skills has also changed and adapted as a result of these disruptive forces. A discussion of the broad themes associated with these forces and the implications that they may have to chemical engineers in the future will be discussed.

542621 Geopolitical Factors Influencing the Evolution of the Chemical Industry

Abstract:

The basic chemicals and plastics industry is built on a simple principle: find a cheap feedstock and convert it into something more valuable. However, the right feedstock depends on place and time. Certain policies and actions of various governments have influenced the evolution and profitability of the chemical industry during the past 50 years, sometimes in surprising ways. For example, high oil prices and US government funding encouraged the development and commercialization of coal gasification technology during the late 1970s and early 1980s. This enabled China to develop a competitive basic chemical industry using coal as a feedstock during the last 25 years. Similarly, an abrupt increase in oil prices during the first decade of the 21st century encouraged the present shale gas revolution, causing a major shift in the economy of producing basic chemicals and polymers in the US. If history is our guide, then what governments choose to fund and regulate may be as important as market forces in determining the shape of our industry and profession in the future.

542622 Agility & Resilience: How to Maintain Career Competitiveness in the Changing Chemical Industry

Abstract:

The world around is changing. The shift to protectionism, global instability and price volatility of raw materials and feedstocks has the potential to wreak havoc on the chemical industry. With the revolution of shale gas in North America and the sudden drop in feedstock costs, the chemical sector in the US has reasserted its historic dominance. Learn how the role of a chemical engineer can contribute to industry growth and resilience, while adapting to allow for personal agility throughout the career life cycle.

527001 Study of Local Boiling Heat Transfer for Micro Nano Surface Structures Using a 3D Transient Heat Conduction Model

Corresponding author. E-mail address: hengyi@mail.sysu.edu.cn (Yi Heng)

Inverse heat conduction problems (IHCP) play an important role in both theory and applications. It has a wide range of applications in different fields, such as aerospace engineering, chemical engineering and nuclear engineering [1-2]. In particular, towards the understanding of a complex nonlinear dynamic process - pool boiling, which is commonly used in chemical engineering, efficient 3D IHCP solution techniques in the study of boiling heat transfer mechanism have received more and more attention. Together with the experimental efforts, the reconstruction of surface heat flux distribution and information about two-phase vapor-liquid structures above the surface are indispensable for full understanding of boiling heat transfer in all stages. To this end, many researchers make every effort to modify surfaces to enhance boiling heat transfer [3]. Wang et al. [4]enhanced heat transfer performance by improving the wettability of surfaces. Enhancing boiling heat transfer by using micropores with reentrant cavities and various microchannel geometries was demonstrated by Deng et al. [5] and Walunj et al. [6], respectively.

In order to precisely study the mechanism of enhanced boiling in micro-nano structures, we consider complex geometric domains (inspired by the micro-nano surface structures produced in our recent work [7]) to accurately modeling the 3D transient heat conduction process in pool boiling experiments. We have already analyzed the forward heat conduction problems for flat-plate surface structures focusing on the study of the influence of heater thickness on the inverse solution quality against perturbed temperature measurements [8]. In this work, we want to extend and use the aforementioned analysis method to investigate how honey-comblike porous structures (cf. Figure 1) enhance boiling heat transfer.

Figure 1 Produced honey-comblike micro porous copper surfaces (adopted from [7])

The 3D transient nonlinear forward heat conduction model defined on the computational domain constrained by the complex micro-nano surface structures is illustrated as follows:

$$\rho(\theta)c_p(\theta)\frac{\partial\theta}{\partial t} = \nabla\cdot(\lambda(\theta)\nabla\theta), \qquad in\ \Omega\times(0,t_f), \qquad (1)$$

$$\theta(\cdot,0) = \theta_0(\cdot), \qquad on\ \Omega, \qquad (2)$$

$$\lambda\frac{\partial\theta}{\partial n} = q_i, \qquad on\ \Gamma H\times(0,t_f), \qquad (3)$$

$$\lambda\frac{\partial\theta}{\partial n} = q_r, \qquad on\ \Gamma R\times(0,t_f), \qquad (4)$$

$$\lambda\frac{\partial\theta}{\partial n} = q_b, \qquad on\ \Gamma B\times(0,t_f), \qquad (5)$$

where Ω represents the heater geometry. The observation time is $[0, t_f]$ and n denotes the outer normal on the boundaries. q_i, q_b and q_r denote the heat fluxes through the heated boundary ΓH, the boiling surface ΓB, and the rest of the boundaries ΓR. ρ, c_p and λ denote the temperature-dependent density, heat capacity and heat conductivity of the material, respectively. θ represents the temperature variable and θ_0 corresponds to the initial temperature.

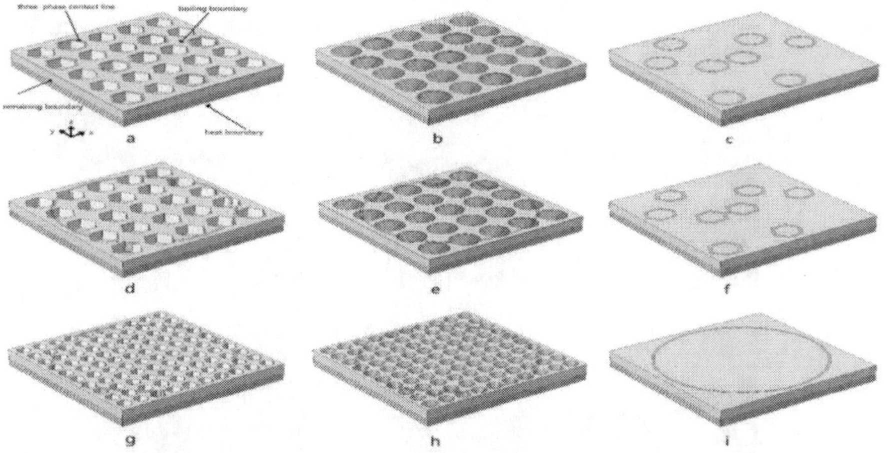

Figure 2 Heater geometries based on the flat-plate surface structure (c, f, i) and two artificial honey-comblike porous structures (hexagonal prisms (a, d, g) and cones (b, e, h)

For the numerical simulations, we consider 9 heaters with the size of 1000um*1000um*130um (see Figure 2). The length of hexagonal micropores are 80um (Figure 2 (a), (d)) and 40um (Figure 2 (g)). The radius of cone micropores are 80um (Figure 2 (b), (e)), 40um (Figure 2 (h)) and the depth of pores are all set to be 75 um. All these geometric settings are based on our recent experiments [7]. Furthermore, according to the micro-layer theory proposed by Stephan [9], most of the heat during boiling is transferred in the micro region of the three-phase contact line (TPCL) by evaporation. Therefore we consider a ring-shaped boiling heat-flux pattern that has been investigated in our previous work [8] for the simulation tests. In Figure 2, the areas of TPCL are marked in blue. The ring width is 16 um, consisting of one part of 8um on the surface and the other part of 8um on the side of hexagonal and cone micropores. The simulated input heat flux q_i at ΓH is set to 0.1 MW/m^2 and the simulation time is $[t_0, t_f] = [0, 0.05]$ s. The heat flux out from TPCL complies with

$$\begin{cases} q_b = 5 \times \sin[100 \times \pi \times (t - 0.02)] + 5MW/m^2 & on\, \Gamma B \times [0.015, 0.03], \quad (1) \\ q_b = 0 & on\, \Gamma B \times (0, 0.015) \cup (0.03, 0.05), \quad (2) \end{cases}$$

Since the surrounding of heater is covered by aerogel blanket we assume q_r=0. The maximum heat flux out from TPCL is set to 10MW/m^2 following the results from our previous work Heng et al. [10]. In this study, we conduct 9 simulations considering different pore structures, pore density and local heat-flux patterns. The initial temperature for all simulations is set to be 143°C. While in the cases of Figure 2(a), (b), (c) and (f) we assume TPCLs randomly occur on the boiling surface that lead to a few small-sized ring-shaped local boiling heat fluxes. A single, large ring-shaped boiling heat-flux pattern is investigated in the cases of Figure 2(d), (e), (g), (h) and (i).

COMSOL Multiphysics is used for the simulations. The finite elements used for all case studies are chosen between 1.6 million and 9 million. The simulation time step is 0.001s. Simulation results of temperature distributions on the heated wall are shown in Figure 3, Table 1 and Table 2, respectively. The order of the results in Figure 3 follows that of the case studies shown in Figure 2. While Figure 3 (a), (b), (c) show the influence of different structures on temperature distributions, Figure 3 (d), (e), (g), (h) show the simulated temperature distributions as the pore density increases.

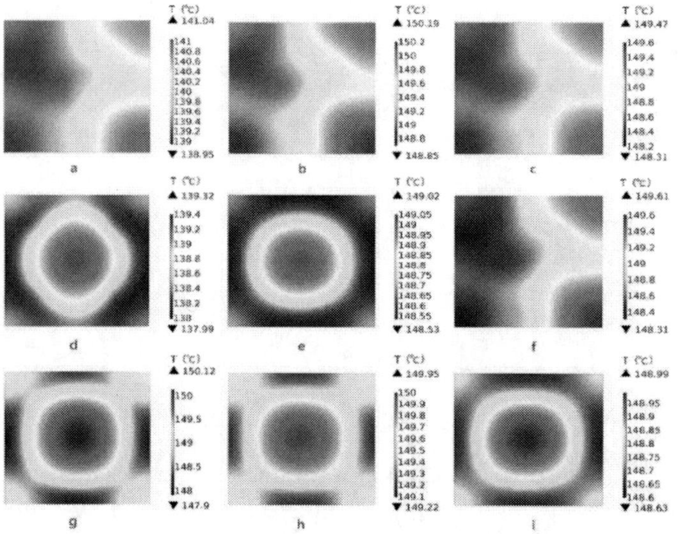

Figure 3 Representative examples of simulated temperature distributions at t = 0.025s

Furthermore, Table 1 and Table 2 summarize the temperature variations on the heated wall in all cases. Due to the increased TPCL areas, at the same level of boiling heat fluxes, the honey-comblike surface structures (Table 1 (a), (b)) lead to higher temperature variations on the heated wall than those on the flat-plate surfaces. As shown in Table 2, we observe that the increasing of pore density for both hexagonal and cone surface structures (Table 2 (g), (h)) lead to even higher temperature variations than those using low-density ones (Table 2 (d), (e)), respectively, which are all higher than those using the flat-plate surface structure (Table 2 (i)). It is concluded that micro-nano surface structures can in general lead to high temperature variations and enhance the local boiling heat transfer. Besides, as discussed in our previous work [8], higher temperature variations in the temperature measurements can also help improve the solution accuracy of IHCP in the presence of same level of measurement error. Hence, the use of micro-nano surface structures shows its advantages.

Table 1 Temperature variations on the heated wall for the simulations of a few randomly distributed small-sized ring-shaped boiling heat fluxes

Temperature(K)	a	b	c	f
Average	412.92	422.56	421.95	421.99
Maximum	414.19	423.34	422.62	422.76
Minimum	412.10	422.00	421.46	421.46
Variations	2.09	1.34	1.16	1.30

Table 2 Temperature variations on the heated wall for the simulations of a single, large ring-shaped boiling heat flux

Temperature(K)	d	e	g	h	i
Average	411.71	421.87	422.16	422.70	421.92
Maximum	412.47	422.17	423.27	423.10	422.14
Minimum	411.14	421.68	421.05	422.37	421.78
Variations	1.33	0.49	2.22	0.73	0.36

In summary, we studied the local boiling heat transfer for micro-nano surface structures using a 3D transient heat conduction model. The effects of selecting different surface structures and pore densities on the temperature distribution results were investigated. Numerical results show the modified micro-nano surface structures can help enhance local boiling heat transfer, which validated our experimental studies [7]. One of our future research will be the study of more complex surface structures containing crystal branches arising from real experiments. Furthermore, we aim to extend our CG-based Iterative regularization method [8] to solve 3D transient IHCP that can deal with complex surface structures. In this way, suitable optimal experimental design techniques are expected to be developed to optimize the surface preparation procedures and boiling experiments to further enhance boiling heat transfer.

REFERENCES

[1] Y. Wang, et al., Effects of anodic oxidation for combustion chamber on heat transfer of the piston in an aero-engine, Experimental Heat Transfer 30(1) (2017) 46-65.

[2] S. Alyokhina, et al., Unsteady heat exchange at the dry spent nuclear fuel storage, Nuclear Engineering and Technology 49(7) (2017) 1457-1462.

[3] D.E. Kim, et al., Review of boiling heat transfer enhancement on micro/nanostructured surfaces, Experimental Thermal and Fluid Science 66 (2015) 173-196.

[4] Y.Q. Wang, et al., Wettability modification to further enhance the pool boiling performance of the micro nano bi-porous copper surface structure, International Journal of Heat and Mass Transfer 119 (2018) 333-342.

[5] D. Deng, et al., Pool boiling heat transfer of porous structures with reentrant cavities, International Journal of Heat and Mass Transfer 99 (2016) 556-568.

[6] A. Walunj, et al., Comparative study of pool boiling heat transfer from various microchannel geometries, Applied Thermal Engineering 128 (2018) 672-683.

[7] Y.Q. Wang, et al., The Two-Layers Composite Structure of Biomimetic Copper Forest and Honeycomblike Porous Structure to Enhance Pool Boiling Performance, 16th International Heat Transfer conference (submitted) (2018) Beijing.

[8] Y. Heng, Mathematical Formulation and Efficient Solution of 3D Inverse Heat Transfer Problems in Pool Boiling, (2011) RWTH Aachen University.

[9] P. Stephan, et al., A new model for nucleate boiling heat transfer, Heat and Mass Transfer 30(2) (1994) 119-125.

[10] Y. Heng, et al., Reconstruction of local heat fluxes in pool boiling experiments along the entire boiling curve from high resolution transient temperature measurements, International Journal of Heat and Mass Transfer 51(21-22) (2008) 5072-5087.

531617 Novel Adsorption-Based Separation Technology for Gas Treating Applications

Abstract:

Gas processing and treating systems often can be expensive, have a large footprint, and require large utility loads and elaborate maintenance programs. With the prevailing low commodity price environment, there is increased focus on reducing costs associated with these systems. This is particularly important when operating in challenging cost environments such as deep water, arctic, or remote offshore. In response to these challenges, ExxonMobil has been developing a novel adsorption-based gas separation technology with a focus on process intensification. This technology uses rapid pressure- and/or temperature-swing cycles to dramatically reduce equipment size and weight, and overall gas treating footprint by ~50%, when compared to conventional technologies. Furthermore, this technology is well suited for a skid-based deployment approach, which has significant execution efficiencies. The weight and footprint savings have several benefits, in addition to CAPEX savings, particularly in locations where space is at a premium.

The current focus of development is deep dehydration and CO_2 removal for cryogenic applications, such as NGL recovery or LNG. A significant amount of fundamental research and pilot testing has been completed demonstrating technical feasibility. This paper will provide background on the technology platform, the key enablers and overall process integration. Furthermore, this paper will also present data from the pilot studies and the approach adopted for scale-up and demonstration.

533415 Process Intensification, a Promising Approach in Separating a Ternary System Using Distillation to Reduce Energy Consumption: Case Study

Abstract:

Separating a ternary mixture into its pure components is a very often required in most of the processes observed in the various industries. At this moment, Distillation is the commonly used process to separate binary or ternary or multi component mixtures provided enough boiling point difference exist among the components. Though it is a frequently used process, distillation is an energy intensive process, which necessities thorough optimization in terms of energy to meet the profitability and sustainability. As the ternary and more component mixtures require series of distillation columns to get the pure components, a systematic approach to effectively optimize the performance of the distillation units in terms of energy, product purity, and operability, is required. For that reason, employing process intensification, energy integration with thorough thermodynamic analysis is a way forward for a cost-effective and energy efficient distillation column network.

In this study, a systematic energy integration of the distillation separation system is developed to get minimum energy consumption with less capital cost. The optimization is carried out in terms of distillation sequence, process intensification, pressure, temperature & composition of the column, no.of stages, and feed tray location. The extent of the energy reduction depends on the composition of the feed as well. The above-explained approach has been utilized in a real time case, which consumes 20% less energy than that of presently employed conventional distillation method for the chosen ternary system. This new method also requires less capital cost for a green field project. The paper also advises general heuristics to employ a particular distillation sequence for separating a ternary mixture system. The methodology employed in the study, can also be utilized to optimize four-component mixture systems.

535443 Improved Fixed-Bed Transport Characteristics: A Shortcut Method to Optimize Catalyst Pellet Specifications

Abstract:

Highly efficient reactors are fundamental for the resource- and energy efficient production of chemicals. The basis to develop such optimal chemical reactors is the simultaneous consideration of all possible design variables, leading to a perfectly tuned system. In catalytic fixed-bed reactors, the catalyst pellet specifications have a significant impact on the system behavior. Besides influencing the catalytic activity and selectivity due to intraparticle transport phenomena, the pellet also influences the fixed-bed transport characteristics such as heat transfer and pressure drop. Hence, the simultaneous design of reactor and catalyst pellet can lead to an improvement of the overall system performance.

For model-based reactor design via dynamic optimization [1] a rigorous treatment of the catalyst pellet during reactor design demands the discretization of the pellet balance equations in addition to the reactor balance equations. The result is a complex heterogeneous model leading to challenges in regard to the computational effort and the numerical stability during optimization. Alternate approaches exist that enable a consideration of the intrapellet mass transport effects via analytical equations. However, these approaches either demand extensive pre-calculations (e.g. Adomian decomposition method [2]) or are strongly limited in regard to their applicability to certain reaction systems or reaction kinetic models (e.g. Thiele-modulus concept [3]). To avoid the drawbacks of these approaches, in this contribution a novel shortcut method is developed and applied. This approach is applicable to arbitrary reaction networks and reaction kinetic models. It enables the computationally cheap optimization of catalyst pellet specifications to improve the bed transport characteristics during the simultaneous design of reactor and catalyst pellet.

The principle of the method is to linearize the rate of formation of each component influencing the reaction rates at bulk conditions. These linear expressions are inserted into the corresponding catalyst pellet component balances. The resulting set of identically structured second order inhomogeneous differential equations is parametrized, decoupled and then solved analytically. The solutions, differing only in the values of their parameters, yield the component concentration profiles in the pellet in dependency of the pellet specifications. Effective reaction rates, calculated via integration, can be inserted into the reactor balance equations. Hence, pellet specifications can be optimized without the need of a heterogeneous model. To ensure a high accuracy of this approach, high catalyst efficiencies should be targeted.

To demonstrate the potential of a simultaneous reactor and catalyst pellet design with the developed shortcut method, a fixed-bed reactor with constant coolant temperature for the production of ethylene oxide is optimized. The objective is to minimize the pressure drop while fulfilling other reactor performance constraints. Pellet specifications in terms of pellet size and pellet heat conductivity are optimized. A pressure drop reduction of 50 % compared to an optimized reactor with prefixed pellet specifications is achieved. Figure 1 shows the optimized catalyst heat conductivity as well as the resulting reaction temperature with reaction progress.

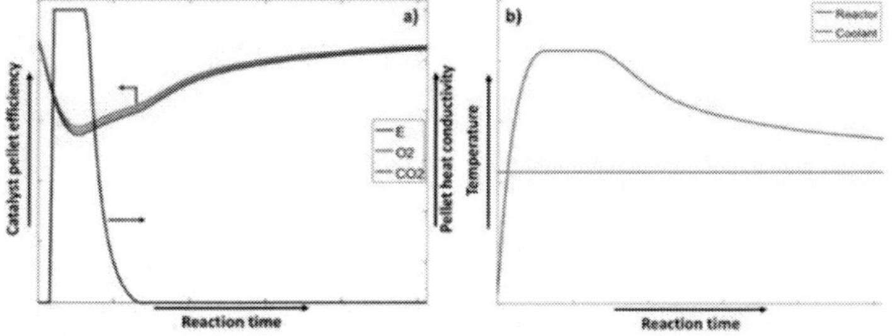

Figure 1: Left: Optimal catalyst pellet heat conductivity profile and resulting catalyst pellet component efficiencies (E=Ethylene, O2=Oxygen, CO2= Carbon dioxide). Right: Resulting reaction temperature profile and optimized coolant temperature.

The presented method enables the design of tailored reactor-catalyst pellet systems to improve the bed transport characteristics. Results with respect to the pellet specifications can point to certain catalyst pellet types available and/or give impulses to develop pellets fulfilling the optimal specifications, leading to highly efficient reactor systems.

The approach developed in this work is generally applicable, i.e., no limitations exist regarding the applicability to certain reaction networks or specific types of reaction rate models as compared to existing methods. At the same time, since a high model complexity is avoided, the approach can be used for efficient simulation and optimization of complex reaction systems.

References

[1] Peschel, A.; Freund, H.; Sundmacher, K.; Ind. Eng. Chem. Res. 49 (2010) 10535-10548

[2] Adomian, G.; J. Math. Anal. Appl. 135 (1988) 501-544

[3] Thiele, E.W.; Ind. Eng. Chem. 31 (1929), 916-920

[4] Aris, R.; Chem. Eng. Sci. 13 (1960) 18-29

537471 Multi-Material, Microchannel Heat Exchanger Design for Enhanced Heat Exchange Processes By Anisotropic Conduction

Abstract:

Recent advances in rapid prototyping techniques, namely laser powder bed fusion, present an opportunity for intensification of heat exchange processes through anisotropic conductive heat flow. Multi-material microchannel heat exchangers â€“ which exploit highly conductive, micro-pillar architecture â€“ are explored here for enhanced rates of heat transfer in the direction of interest only. By reducing heat flow in undesired directions and decreasing the characteristic time for heat transfer between the fluids, it is shown that a reduction of residence times and heat-exchanger volume can be achieved. For a variety of cases, volumetric heat transfer coefficients for systems with and without conductive micro-pillars are evaluated numerically and compared. A 3D printed prototype is also built and tested to demonstrate the functionality of multi-material heat exchangers and to verify numerical results.

538723 Small-Scale Production of Platform Chemicals from Coal with Low-Temperature Microwave Plasma

Abstract:

The United States is endowed with extraordinary riches: it holds more than a quarter of the world's coal reserves. Regrettably, this wealth is severely undervalued. Although coal has catalyzed industrial development in the 19th and early 20th century not only as a ready source of energy, but also as the principal source of synthetic chemicals and materials â€" from aspirin to benzene to naphthalene to tarmac, â€" over the past decades coal became essentially synonymous with production of low-cost electricity.

Shale revolution has resulted in an abundance of low-cost natural gas and prompted the fall of coalâ€™s share in electricity production from 50% to less than 30% over the course of a single decade. With significantly depressed domestic thermal coal prices, there is an opportunity to reorient the chemical industry back to reliance on this accessible and abundant feedstock. It has already happened in China, for instance, where the present capacity of coal gasification/Fischer-Tropsch conversion to methanol targeting production of olefins is 13.6Mtpa, with $55 billion worth of new capacity planned through 2026. FT plants, however, have conventionally been associated with outsized CAPEX investment (in billions USD) and a large environmental footprint, precluding their adoption in the United States.

A viable alternative method for production of coal-derived platform chemicals is direct, thermal conversion (pyrolysis) of coal. Known and industrially employed since 1800s, destructive distillation of coal yields hydrocarbon gases, liquids (monoaromatics and coal tar), and coke. In fact, until 1950s, the coke ovens were responsible for virtually 100% of the domestic benzene supply, but were quickly displaced by petroleum refiners. While the first benzene produced from petroleum appeared only in 1950, by 1961 petroleum benzene accounted for more than 75% of the total production.

Today, benzene and other monoaromatics (BTEX) are the crude oil-derived platform chemicals at the base of many important industrial supply chains: plastics, rubbers, fabrics, solvents, detergents, dyes, and pharmaceuticals. The global demand for benzene alone is in the order of 40 million tons per year. Domestic benzene deficit is expected to grow from 1.8 million tons in 2014 to 4.4 million tons by 2023, due to impact of the shale revolution on domestic BTEX sources. Refineries are forced to run reformers at reduced rates, while BTEX streams from naphtha cracking are reduced as low-cost high-ethane shale gas becomes preferred feedstock for ethylene production.

H Quest has developed and demonstrated a novel coal conversion process that avoids heat transfer and scale requirements limitations of conventional coal conversion technologies, and is well-suited for small-scale, distributed production of chemicals and material precursors from coal. In place of conventional thermal inputs, this rapid, continuous process uses focused microwave energy to create non-equilibrium plasma within inert or hydrogen-rich (CH_4, H_2 or syngas) process gas, localized around entrained coal particles. Optical (spectrometry) diagnostics are used for reaction characterization, species identification, and temperature determination. Observed intensity ratios or spectra band shapes yield temperature by Boltzman analysis. Pyrometry measurements of the solid particles falling through the MW reaction zone yield a temperature of ~ 1200 K. The entraining gas remains relatively cool: thermocouple measurements at the outlet of the microwave reactor typically do not exceed 300 C. Thus, rapid (< 1 sec) pyrolysis of coal releases liquid products and intermediates into the relatively cool process gas, where immediate quenching prevents secondary cracking reactions. The result are the high yields of the relatively high-value liquids (50-60wt%, d.a.f) and minimal yields of the low-value gases (<5wt%).

Process gas composed of methane or syngas serves as a hydrogen source within the single-stage reactor, eliminating external hydrogen production units and the associated CO_2 production, water consumption, and capital costs. Its ionization within the localized reaction zones produces active CH_x* and H* radical species, which promote further coal decomposition, prevent retrogressive radical recombination reactions of oil intermediates, and improve quality of the liquid products through partial hydrogenation and alkylation reactions. Direct activation of a hydrogen-rich gas within localized reaction zones eliminates requirements for high pressures, which conventional coal-to-liquids processes relies on to overcome mass transfer limitations.

Thus, the process enables deployment of a continuous, small-footprint, high-throughput conversion system with *a thermal efficiency approaching 80% and life-cycle CO_2 emissions and water consumption at least 50% lower than conventional coal conversion technologies.*

Varying energy input, feed rate, and process gas composition allows adjusting the liquid product composition and properties. Bituminous coals processed in inert atmosphere (process gas includes inert, recycleable argon) yield highly aromatic liquids. Up to 25% of the liquids are benzene and other monoaromatics. Co-processing with activated hydrogen-donating gases results in methylation, hydrogenation, and ring-opening of the aromatic compounds, as shown by GC/MS traces of the resulting products and corroborated by optical spectrometry. Experimental data from testing high-sulfur (>3%) bituminous coals from Illinois and Ohio shows reduction of sulfur content relative to the parent coal in both char and liquid products of 50% and 80% respectively, effectively beneficiating both products. Production of excess hydrogen from methane decomposition may be used in downstream processes or sold as a valuable by-product.

In the past, coal conversion processes were hindered by inefficiencies, vast capital costs, high environmental impact, and, as a result, by non-competitive costs of production. New technologies relying on unconventional pathways and methods can enable cost-effective, small-scale, clean, efficient value-added production of high-value chemicals and materials directly at the coal mine sites. Most importantly, deployment of these technologies would have a direct and rapid impact on job creation and economic development in the rural communities most affected by the recent downturn in the coal industry.

Acknowledgements

H Quest Vanguard, Inc. is a privately held technology company, based in Pittsburgh, Pennsylvania, focused on the development and commercialization of novel hydrocarbon conversion technologies. Work supported by the Department of Energy, Office of Science under award DE-SC0015895 and by the Illinois Clean Coal Institute under subgrant 15-01 have contributed to materials presented herein.

538809 Effectiveness Factor Phenomena for the Transition between PBR (Packed Bed Reactor) and MR (Membrane Reactor) Via Coupled Heat and Mass Transfer

Abstract:

The study of multiphase transport/reaction phenomena is, of course, of interest to a wide range of separation and reactor applications involving of simultaneous heat, momentum, and multi-component mass transport. The most marked characteristic of such multiphase systems is their nonlinear nature, which requires the quantification of a multiple-coupled set of nonlinear differential equations for the problem solutions that entails a complex set of technical challenges. For reactors, including the MR studied here, accounting for all such phenomena dictates the use of the so-called heterogeneous models rather than the simplified pseudo-homogeneous models that have, to date, dominated the MR modeling literature. This work focuses on the multi-scale modeling of a membrane reactor (MR within which the water gas shift (WGS) reaction is carried out. The work assesses the MRâ€™s effectiveness in transforming, coal-derived syngas into two streams, one rich in hydrogen and the other rich in carbon dioxide. It is shown that significant variation of the catalyst pellet effectiveness factor occurs along the reactorsâ€™ axial direction, and that catalyst pellets of the same diameter exhibit different effectiveness factors within the PBR and MR environments. The effect of catalyst solid thermal conductivity on temperature gradients within the catalytic pellet, and the effect of sweeping gas pressure/temperature on the MRâ€™s behavior, are both quantified. Adiabatic and wall-isothermal operation are shown to exhibit significant differences from each other for both PBR and MR reactors.

527461 Carbon Dioxide Introduced for Green Synthesis of Thiuram Disulfides Vulcanization Accelerators Under Mild Conditions

Abstract:

Thiuram disulfides have been widely used in rubber industry as rubber vulcanization ultra-accelerators. The conventional industrial process for the production of thiuram disulfide included reacting a secondary amine with carbon disulfide in an aqueous solution of sodium hydroxide to form a sodium salt of dialkyldithiocarbamate acid, then adding the mixed solution of sulfuric acid and hydrogen peroxide to neutralize and oxidize sodium dialkyldithiocarbamate. The precipitate, thiuram disulfide, can be obtained by filtering, washing, drying. While this method uses the large quantities of sulfuric acid and generates huge amount of salt by-product, sodium sulfate. Thus, an eco-friendly method for oxidation of sodium dialkyl dithiocarbamate to the corresponding thiuram disulfides has been developed using carbon dioxide as neutralizing agent in the presence of H_2O_2 as the oxidant and water as the solvent. After optimizing the reaction conditions such as temperature, concentration of reactants, flow rate of carbon dioxide, dropping speed of hydrogen peroxide, aging time and so on, this method could provide green mild reaction conditions, reduce CO_2 emission and H_2SO_4 consumption remarkably, and reach the almost closely high yields and purity of products as traditional method. With the HPLC-MS analysis of mother liquor, it is proved that the main side reaction is the peroxidation of dialkyldithiocarbamate acid which produced corresponding sulfoxide and sulfone. By comparing the alkalinity of sodium carbonate, sodium bicarbonate and sodium dialkyldithiocarbamate, doing related experiments and reading relevant references, the reaction path has been proposed.

530545 Plastic Waste Accumulation Problem and Emerging Solutions

Abstract:

More than eight billion tons of plastic waste has been accumulated over the past 50 years. The majority (90%) of the worldâ€™s plastic waste goes directly into landfills and 3% ends up in the oceans. At the current rate, the planet will have 30 billion tons of plastic waste by 2050, and the oceans will have 250 million tons of plastics by 2025, or â€œone pound of plastics for every three pounds of fishâ€. Plastics are persistent in environment and degrade very slowly (~100 years timescale), while releasing toxic microplastics and chemicals into the landfills and oceans. This poses serious threats to our ecosystems, drinking water, and food supply. Our analysis shows that to remove the plastic waste from landfills is estimated to cost about $0.5/kg, or $4 trillion globally. To remove the plastic waste from the oceans using the state of the art technology is estimated to cost about $\$10^{18}$, about 10,000 times the global GDP.

Polyolefin waste accounts for about 63% of the total accumulated plastic waste. About $100 billion of post-consumer polyolefin products are consigned to landfills annually. Converting polyolefin waste into useful products can reduce the accumulated waste and associated risk to the environment. In this study we compare incineration, which is the most mature technology of reducing plastic waste, with two emerging solutions, converting polyolefin waste into (1) pristine polymers or (2) clean fuels. The results show that both solutions potentially are more profitable than incineration. Sorted polyolefin waste can be converted into pristine polymers with the highest potential profit. Mixed polyolefin waste can be converted into clean fuels with the second highest profit. The supplies of the sorted waste, however, are limited presently. Developing more efficient sorting technologies or changing the current waste management system to collect more sorted polyolefin waste should have a major positive impact on reducing the polyolefin waste accumulation.

Process Systems Engineering: Limits of performance of the "cyclic" waste plastic economy

James Alistair Fox, Baraka Celestin Sempuga

Institute for the Development of Energy for African Sustainability, University of South Africa, Florida Science Campus, Johannesburg, South Africa

Abstract

The rapid accumulation of waste plastics in the environment has led to recent media and public attention focused on an impending environmental crisis. One potential solution is the idea of a "cyclic plastic economy". It should be possible to maintain an entirely closed and independent cycle, with negligible environmental impact. Old plastics enter the cycle while new plastics leave the cycle. How realistic is this, though? Just how "cyclic" is the cycle? This work seeks to investigate this question by applying high-level systems engineering to identify the limits of performance for such a cycle.

It was found that, if conventional fossil fuel based power generation methods were used, the cyclic economy would have a maximum carbon efficiency of 97.5%. While this is a very high efficiency, for the cycle to be completely closed, a carbon efficiency of 100% is required. An efficiency of 100% can be achieved if renewable energy sources are used instead.

While a real process will always perform worse than this idealized result, the analysis shows that a cyclic economy will never be feasible if non-renewable energy is used. However, it can be conceptually feasible with the use of renewable energy.

Introduction

The plastics industry has seen consistent growth for many decades, with the global value of the industry predicted to reach USD 654-million by 2020[1]. This growth has been accompanied with significant growth in waste generation. It is estimated that some 8-million tons of plastic are disposed of in the oceans every year[2]. With waste plastics taking anything from 100 to 1000 years to degrade naturally, plastic waste is rapidly accumulating in the environment.

China had been importing as much as 51% [3] of the worlds waste plastics, leading many nations and economies to adopt an "out of sight-out of mind" view on their waste production. However, China ceased their imports at the end of 2017, leaving those same economies with vast amounts of waste plastic to deal with and garnering significant media and public attention. This has led to an upsurge in new research in how to deal with plastic waste.

Much of the attention has been focused on the idea of a "cyclic plastic economy"; the principle of this idea is that plastics are first mechanically recycled (washed, repaired and reused) as much as possible until their condition has deteriorated to a point where the plastic needs to be chemically broken into their chemical intermediates. This transformation is primarily accomplished using pyrolysis, which breaks the plastics down to alkene molecules. These alkenes can then be re-polymerized into new plastic products.

In principle, it should be possible to maintain an entirely closed and independent cycle, with negligible environmental impact. Old plastics enter the cycle while new plastics leave the cycle. How realistic is this, though? Just how "cyclic" is the cycle? This work seeks to investigate this question by applying high-level systems engineering to identify the limits of performance for such a cycle.

The proposed system to investigate this question is comprised of two major sections:

1. Pyrolysis, which is an endothermic reaction that involves using heat to split the polymer chains into predominantly alkene components that is typically carried out at temperatures in the region of 823K.
2. Polymerization, which is an exothermic process that is carried out at temperatures around 373K. Normally, coordination polymerization of ethylene monomers is used. However, pyrolysis does not produce ethylene alone but rather a range of alkene products. The re-polymerization of pyrolysis products for this system would require reversing the pyrolysis reaction using something analogous to radical polymerization. However, for this work, the mechanism of polymerization is not important but only that it is the reverse of pyrolysis.

Idealised thermodynamics for the cyclic economy

In order to begin an analysis of the hypothetical cyclic economy it is necessary to represent the processes involved with their respective material balances.

Due to limiting hydrogen, pyrolysis primarily produces unsaturated alkene hydrocarbons of varying carbon chain lengths. The relative amounts of each alkene depends on the temperature at which the pyrolysis is carried out. Most empirical research [4–6] carried out on plastic pyrolysis reports the product distribution in mass fractions of gases, liquid fraction and heavy hydrocarbons. At a temperature of 873K, a typical pyrolysis yield is given as 25% gases, 45% liquids and 30% heavies.

For the purposes of a complete material balance, it would be better to have a full distribution of all the alkene products but such detailed information is not available in the literature. Instead, each fraction will be represented by a "representative component", the component that comprises a significant proportion of a given fraction. The gas fraction will be represented by propylene, the liquid fraction by heptadecene and the heavy fraction by triacontene. With the given information it is now possible to write the material balance for polyethylene pyrolysis.

$$(C_2H_4)_n \rightarrow 0.17nC_3H_6 + 0.05nC_{17}H_{34} + 0.02nC_{30}H_{60}$$

Here "n" is the number of ethylene monomers present in a single mole of polyethylene. Since this value can vary over a wide range, the monomer number will be treated as an unknown variable. It is important to note that this does not represent a reaction for pyrolysis; it represents a material balance for pyrolysis as a whole. In addition, this material only applies to pyrolysis conducted at 823K. Investigation of a different temperature would require data on the product distribution at that specific temperature. With sparse data on plastic pyrolysis available, this work will stick to using the data at 823K.

The goal of the cyclic economy would be to re-polymerize the alkenes back into polyethylene; this would need to be done perfectly to ensure that the loop is closed. The mechanism for the re-polymerization would probably need to be something akin to radical polymerization. This analysis will not be investigating the mechanism but rather the system as a whole. The material balance for re-polymerization would be represented as the reverse of the material balance for pyrolysis.

Returning to the first step of the system, pyrolysis, the limit of performance would be reversible pyrolysis. A hypothetical pyrolysis process that would operate reversibly is represented in Figure 1

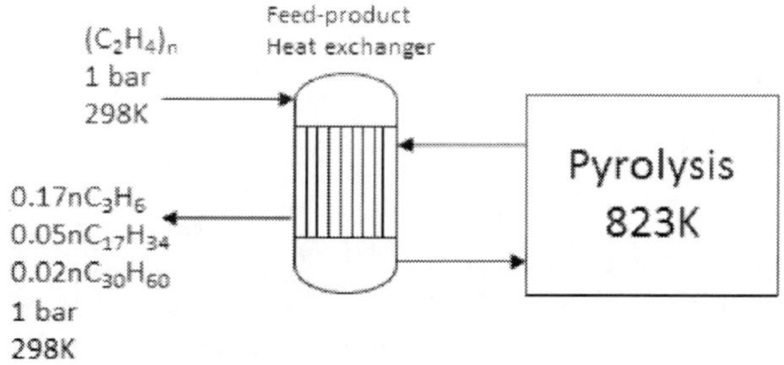

Figure 1: Representation of a reversible pyrolysis process

The process in Figure 1 is not a practical one. One of the most obvious problems is that the feed-product heat exchanger would need to have an infinite surface area. The purpose of this idealized analysis is to find the limit of performance and provide a basis for comparison. No real process can achieve what is represented in Figure 1 but if the cyclic economy does not work under idealized conditions, it will *never* work under realistic conditions and that applies for all systems.

Applying the first law of thermodynamics to Figure 1 reveals that, for every mole of polyethylene, pyrolysis needs 72.6n kW of heat supplied at a temperature of 823K. This energy needs to come from somewhere. One of the major premises of the cyclic economy is that this energy can come from the exothermic re-polymerization. This is simply the process in Figure 1 with the arrows reversed but there is one important difference, the polymerization takes place at a lower temperature (around 373K) than pyrolysis. This immediately implies a violation of the second law of thermodynamics, since the heat produced in the re-polymerization would need to transfer to pyrolysis against the temperature gradient, but it is possible to move heat against the gradient with heat pumps.

The goal is now to devise a method of "upgrading" the heat produced by re-polymerization and then transfer that heat to pyrolysis. Consider the ideal/reversible re-polymerization process, Figure 2

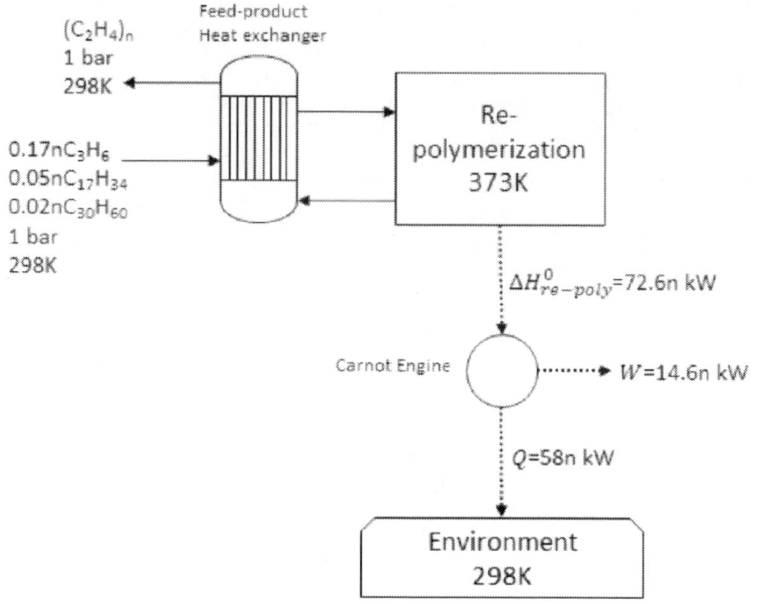

Figure 2: Ideal re-polymerization with work recovery using a Carnot engine. Solid arrows indicate mass flows; dotted arrows indicate energy flow

In Figure 2, the heat generated by the re-polymerization is sent to a Carnot engine where work is recovered by the engine and remaining heat expelled to the surroundings at ambient conditions. The well-known Carnot's Theorem calculates for the work recovery

$$W = Q_h\left(1 - \frac{298K}{T}\right)$$

Where:

- W is the work recovered by the engine
- Q_h is the heat entering the engine from the hot reservoir. In this case, $Q_h = \Delta H$.
- T is the temperature of the hot reservoir.

The re-polymerization is able to provide 14.6n kW of work towards the upgrading of energy.

This same idea can then be applied to the process in Figure 1,

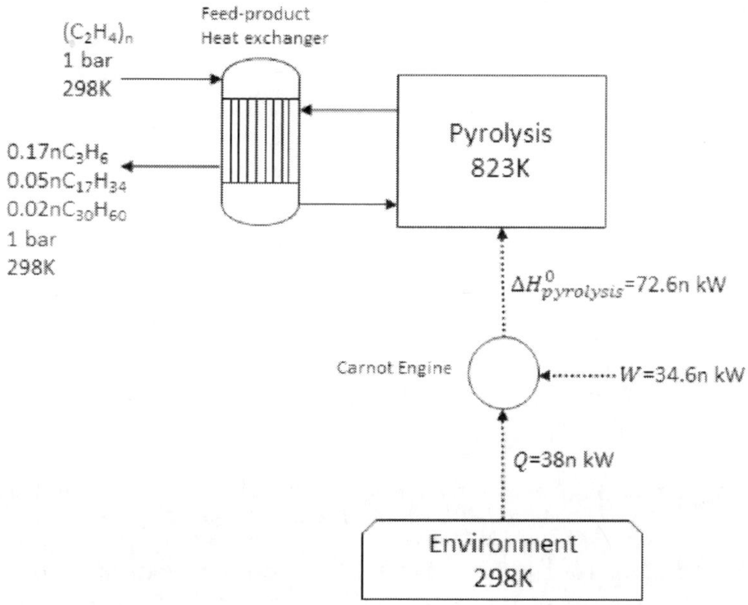

Figure 3: Ideal pyrolysis with work addition using a Carnot heat pump. Solid arrows indicate mass flows; dotted arrows indicate energy flows

Figure 3 shows that 34.6n kW of work is needed in order to upgrade the heat from the environment, but the re-polymerization is only capable of supplying 14.6n kW. It is clear that there is a deficit of 24n kW of work that needs to be supplied from an external source.

This result already shows that the cyclic economy is not a completely closed cycle, even under ideal circumstances. The question is now: how far off is the cycle?

In order to complete the construction of the idealized chemical waste plastic recycling process, a source of energy is needed to supply the outstanding 24n kW of work. The easiest way to do this is to use electricity. The material balances for how this electricity is produced then need to be taken into account. Previous work by one of the authors involves the design of a process that can convert waste plastic into electricity. The initial estimates show that waste plastic could produce up to 34 GJ of power per ton of waste polyethylene. The material balance that allows this process to provide 24n kW of electricity is given by:

$$0.025(C_2H_4)_n + 2nAir \rightarrow 0.05nCO_2 + 0.1nH_2O + 1.94nAir$$

A portion of the waste polyethylene now needs to be diverted to a process that can convert the polyethylene into electricity. In addition, this process now produces carbon dioxide, a different type of pollutant. However, it is now possible to re-polymerize the waste polyethylene into new "virgin" polyethylene.

It is now possible to construct a diagram by combining Figure 2, Figure 3 and the electricity production.

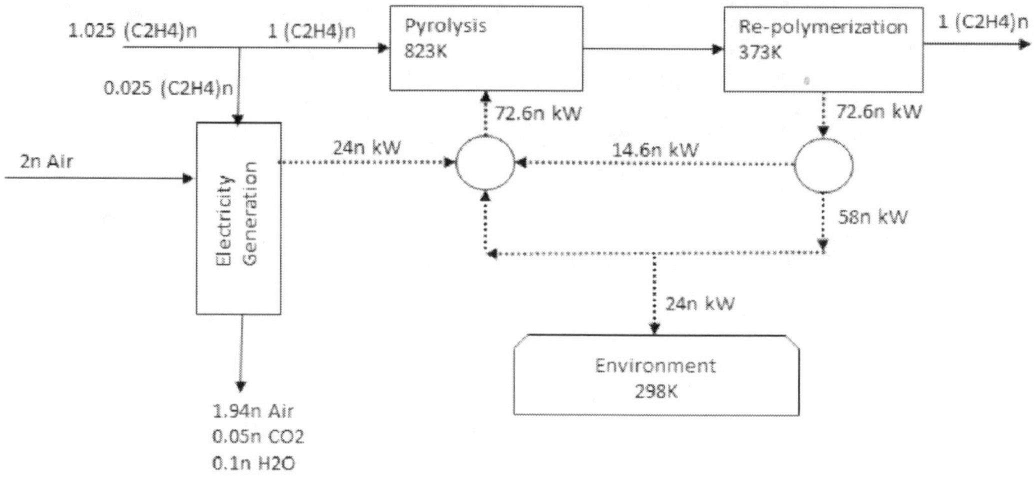

Figure 4: Supplying the energy deficit through electricity via "polyethylene to power" process. Solid arrows indicate mass flow and dotted arrows indicate energy flows.

The system shown in Figure 4 is now overall exothermic. Additional work (in the form of electricity) is now being added to account for the work deficit. In order to maintain the conservation of energy, an identical amount of heat now needs to be rejected from the system. However, this heat has the same temperature as the surroundings and thus no potential is being lost.

To assess the efficiency of the system, or how close to "closed" the cycle is, the carbon efficiency will be used. The carbon efficiency is defined as the amount of carbon in the desired products, relative to the amount of carbon in the feed.

$$C_{eff} = \frac{C_{products}}{C_{feed}} = \frac{2\ mol/s}{2.05\ mol/s} = 0.975$$

Where:

- C_{eff} is the carbon efficiency
- $C_{products}$ is the amount of carbon in the desired product. Here, the desired product is polyethylene. Carbon dioxide is not a desired product.
- C_{feed} is the amount of carbon in the feed.

While 97.5% is a very high carbon efficiency, a carbon efficiency of 100% is required in order for the system to be completely cyclic. In addition, this efficiency is achieved by generating electricity from waste polyethylene, which is not a widely used technology. If this electricity was instead provided by a more typical fossil fuel, like coal, the efficiency of the system would be even lower.

This shows that the cyclic economy will never be an entirely closed and sustainable system unless some way can be found to provide the energy deficit by some method that does not result in the emission of other types of pollutant. This leads to the possibility of using renewable energy.

A closed cyclic economy loop using renewable energy

Wind and solar power offers an opportunity to supply the work deficit of the waste plastic recycling system without emissions that will reduce the efficiency of the system.

The costs of renewable energy installations are constantly changing and have a strong dependence on the region and current technology, but overnight costs can be approximate from data available in the literature [7]. An industrial wind turbine rated at 1.5MW would have an overnight cost of approximately 2.5-million USD. The amount of waste plastic that could be chemically recycled by this turbine is given by,

$$Recycled = \frac{P_{rated}\mu}{W_{deficit}} . M_{C_2H_4}$$

Where:

- Recycled is the mass of waste plastic that can be chemically recycled
- P_{rated} is the rated power of the renewable energy source. 1500 kW for the wind turbine/
- μ is the efficiency of the wind turbine, assumed to be an average of 30%
- $W_{deficit}$ is the work deficit required by the chemical recycling system, 24n kW
- M_{C2H4} is the molar mass of polyethylene, 0.028n kg/mol

The amount of waste polyethylene that could be chemically recycled by a 2.5-million USD, 1.5 MW wind turbine thus comes to 0.525kg/s or 45.4 tons/day.

Using the same reasoning a 1000 kW Concentrated Solar Power (CSP) array, at a cost of 6300 USD/kW [8] and an assumed efficiency of 40%, could chemically recycle 40 tons/day of waste polyethylene.

Taking the market value of High-density polyethylene as around USD 670/ton and assuming the chemically recycled plastics could be sold at the same value, a rough payback period for a renewable energy installation would be between 85 and 160 days. This number only considers the overnight cost of the renewable energy itself, with no operating expenses or the capital for the other units. Even so, such a payback period is not unreasonable.

The system could now be represented as,

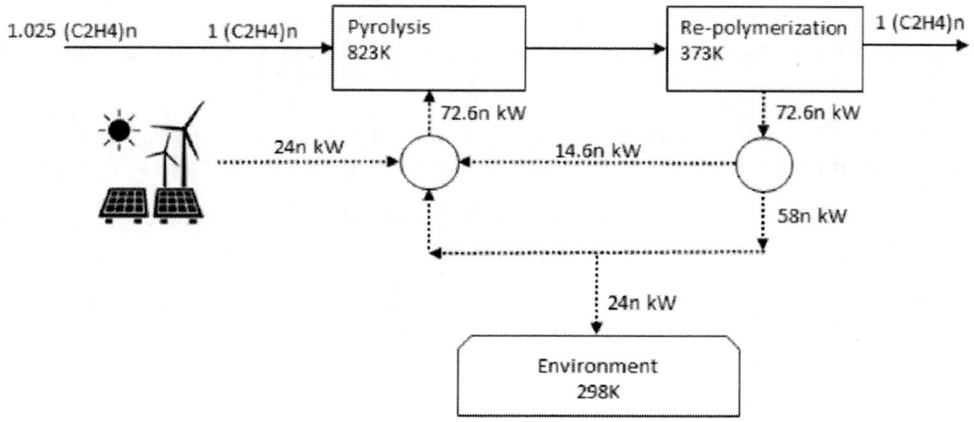

Figure 5: Conceptual closed loop chemical waste plastic recycling system

Instead of using a carbon dioxide emitting energy source, the system in Figure 5 gains the 24n kW of work deficit from renewable energy. This system does achieve the required carbon efficiency of 100%. This shows that, provided a renewable energy source or non-emitting energy source is available, the "cyclic economy" is expensive but conceptually viable.

Conclusion

Conventional power production invariably requires some form of combustion but the use of renewable energy avoids this issue and presents the possibility of achieving the required 100% carbon efficiency needed for a truly cyclic system. It was found that a 1.5 MW wind turbine, operating at 30% efficiency could provide the energy requirements for a cyclic system processing 45 tons/day of waste polyethylene. A 1000 kW concentrated solar power installation, operating at 40% efficiency, could process 40 tons/day. These renewable energy sources carry substantial capital costs of between 2.5 and 6 million US-dollars, for the renewable energy units alone. However, they do achieve the required 100% carbon efficiency needed for the cyclic system.

All of these values represent an idealized limit of performance, the very highest efficiency any process could hope to achieve. Real processes would always perform worse than this limit, general energy loses and fluctuations in renewable energy power supply are just some of inefficiencies present in real systems.

The implication posed by this analysis is that polymers contain a significant amount of chemical potential locked away in the polymer structure. There could be great potential in viewing waste plastics not as a "waste" but rather as an opportunity, a potential new fuel source waiting to be unlocked.

References

1. Grand View Research: Plastics Market Size Analysis, Trends | Global Industry Report, 2020.
2. Jambeck, J.R., Geyer, R., Wilcox, C., Siegler, T.R., Perryman, M., Andrady, A., Narayan, R., Law, K.L.: Marine pollution. Plastic waste inputs from land into the ocean. Science. 347, 768–71 (2015). doi:10.1126/science.1260352
3. Reuters: China's plastic demand to rise as foreign garbage ban to curb recycled supply, https://www.reuters.com/article/us-china-plastics-demand/chinas-plastic-demand-to-rise-as-foreign-garbage-ban-to-curb-recycled-supply-idUSKCN1BO0J8
4. Almeida, D., Marques, M. de F., Almeida, D., Marques, M. de F.: Thermal and catalytic pyrolysis of plastic waste. Polímeros. 26, 44–51 (2016). doi:10.1590/0104-1428.2100
5. Kumar, S., Singh, R.K.: Thermolysis of High-Density Polyethylene to Petroleum Products. J. Pet. Eng. 2013, 1–7 (2013). doi:10.1155/2013/987568
6. Yan, G., Jing, X., Wen, H., Xiang, S.: Thermal Cracking of Virgin and Waste Plastics of PP and LDPE in a Semibatch Reactor under Atmospheric Pressure. Energy & Fuels. 29, 2289–2298 (2015). doi:10.1021/ef502919f
7. Energy Information Administration, U.: Cost and Performance Characteristics of New Generating Technologies, Annual Energy Outlook 2018. (2012)

8. Renewable Energy Agency, I.: RENEWABLE ENERGY TECHNOLOGIES: COST ANALYSIS SERIES Concentrating Solar Power Volume 1: Power Sector Issue 2/5 Acknowledgement. (2012)

528218 Discovery of Electronics Cooling Fluids

Abstract:

Modern miniaturization trends have enabled the development of power intensive electronic devices and resulted in a dramatic increase in electronics cooling requirements that traditional cooling approaches are unable to supply. Efforts to identify working fluids for electronics equipment could enhance cooling performance and meet environmental and safety requirements. Among all cooling approaches, two-phase cooling systems can achieve high heat flux as they rely on phase change of the liquid. Micro-channel heat sinks have become one of the most promising two-phase cooling systems as they allow the combined action of heat conduction, latent heat absorption, and forced convection, leading to sufficient heat dissipation while maintaining a safe device temperature. However, heat transfer performance of micro-channel heat sinks has been plagued by the poor heat transfer characteristics of available cooling fluids. In this work, we aim to identify an efficient replacement of the common industrial cooling liquid HFE-7200.

We build on molecular design methodology [1] that addresses the challenge of large combinatorial search spaces in computer-aided molecular design through decomposition. The approach decomposes the complex CAMD problems into three subproblems: composition design, structure generation, and extended design. This three-phase approach has been implemented in the software AMODEO, which is an ideal tool to solve the coolant design problem. We looked closely into the physical and chemical property constraints that new fluids should satisfy. We used group contribution methods as the major property estimation tool along with additional accurate property models. The resulting candidate molecules were ranked by a heat transfer per unit pressure drop metric, developed specifically for the micro-channel heat sinks regime. Five organic families with the least environmental impact were identified, including hydrofluoroethers (HFEs), hydrofluoroolefins (HFOs), fluorinated esters, ethers, and aliphatics. Most candidate molecules are novel compounds, suggesting promising directions for developing safe, biodegradable, low ozone depleting, and low global warming potential coolants.

[1] A. Samudra and N. V. Sahinidis., Optimization-based framework for computer-aided molecular design., AIChE Journal, 59:3686â€“3701, 2013

535465 Computer-Aided Design of Products Derived from Biomass Pyrolysis

Abstract:

The use of biomass to produce fuels and chemicals required for the modern world has emerged as a promising alternative to reduce our dependency on the fossil fuels which are used as the main feedstocks for the petrochemical industry. Among chemical biomass-to-fuel conversion processes, pyrolysis is considered as the cheapest conversion pathway towards liquid biofuels and it is gaining increased attention as a sustainable option for the production of bio-oil and biochemicals (Galadima and Muraza, 2015). In addition to its significant economic potential, pyrolysis technology has a high degree of flexibility as any type of biomass can be potentially used as a feedstock. Despite its advantages, the use of biomass pyrolysis to produce renewable fuels poses several challenges. The difficult characterization of highly diverse biomass resources, reaction intermediates and final products makes the commercialization of the process quite challenging. Although extensive work has been done on different pretreatment technologies (Hosseini et al., 2010) and reaction pathways of several feedstocks (Papari and Hawboldt, 2015), the upgrading of pyrolysis products into a viable product mix remains a challenge that requires further investigation.

In a recent study, Sharifzadeh et al. (2017) addressed the challenges of modeling the hydrothermal upgrading of pyrolysis oil in order to deoxygenate it and make it usable in the current energy infrastructure. In this work we develop a computer-based methodology for identifying an optimal product mix given information on the effluent from a pyrolyzer. A set of possible products is specified based on their properties and prices, and we formulate an optimization problem to identify the most valuable product portfolio. The proposed methodology builds on a general computer-aided model for designing mixtures and product blends, in which the number, identity and composition of different chemicals in the final product are optimized simultaneously using advanced optimization techniques (Jonuzaj et al., 2016, 2018). The modeling approach is demonstrated though the determination of optimal product blends from predefined feedstock.

References

Galadima, A., Muraza, O., 2015. In situ fast pyrolysis of biomass with zeolite catalysts for bioaromatics/gasoline production: A review. *Energy Conversion and Management 105*, 338–354.

Hosseini, S.A., Lambert, R., Kucherenko, S., Shah, N., 2010. Multiscale modeling of hydrothermal pretreatment: from hemicellulose hydrolysis to biomass size optimization. *Energy Fuels 24*, 4673–4380.

Jonuzaj, S., Akula, P.T., Kleniati, P.M., Adjiman, C.S., 2016. The formulation of optimal mixtures with Generalized Disjunctive Programming: A solvent design case study. *AIChE Journal 62*, 1616–1633.

Jonuzaj, S., Gupta, A., Adjiman, C.S., 2018. The design of optimal mixtures from atom groups using Generalized Disjunctive Programming. *Computers & Chemical Engineering.* URL: *https://doi.org/10.1016/j.compchemeng.2018.01.016.*

Papari, S., Hawboldt, K., 2015. A review on the pyrolysis of woody biomass to bio-oil: Focus on kinetic models. *Renewable and Sustainable Energy Reviews 52*, 1580–1595.

Sharifzadeh, M., Richard, C.J., Shah, N., 2017. Modelling the kinetics of pyrolysis oil hydrothermal upgrading based on the connectivity of oxygen atoms, quantified by ^{31}P-NMR. *Biomass and Bioenergy 98*, 272–290.

524516 Reactive Distillation Combined with Pervaporation for Biobased By-Product Recovery

Abstract:

In downstream processing the biobased industry often struggles with complex multicomponent mixtures. By-products in effluent with low concentration may complicate the wastewater treatment and limit complete usage of biomass. In pulping, an effluent containing mainly acetic acid and formic acid is produced from black liquor evaporation. State of the art process for effluent treatment is reactive extraction followed by several distillation steps for solvent recovery and product isolation. Reactive extraction combined with alternative separation methods were investigated to reduce the high energy consumption and apparatus demand for solvent regeneration and isolation of acetic acid and formic acid from black liquor evaporation condensate. The aqueous mixture of acetic acid and formic acid forms inseparable azeotropes.

Reactive distillation and pervaporation are perfectly suited for separation of azeotropic mixtures. Reactive distillation overcomes azeotropes by changing the substance properties via chemical conversion. With carefully chosen esterification the market value of the by-products may be increased, and the significant change of the physico-chemical properties and product separation decrease the cost of product isolation. For separation of acetic acid, formic acid and water reactive distillation with methanol as esterifying agent was successfully performed, but with low reaction rates without catalyst. For successful solvent regeneration high reaction rates and high separation efficiency of the carboxylic acids are needed. The reaction rates were raised by admixture of 5 wt.% 4-DBSA the catalyst 4-dodecylbenzenesulfonic acid (4-DBSA) to the solvent. Batch reactive distillation experiments of laden solvent showed complete conversion of formic acid and a methyl acetate yield of 88 % in the distillate. The feasibility of catalyst admixture to the solvent phase was successfully tested. The catalyst has a minor impact on the liquid-liquid equilibrium. The distillate contains the low boiling methyl esters and excess methanol. Water is not transferred to the distillate. This product mixture exhibits a low boiling azeotrope of methyl acetate and methanol.

Pervaporation is able to isolate azeotropic mixtures because the separation does not depend on the vapor-liquid equilibrium; it is the result from solution-diffusion properties of the components in the membrane. Therefore, pervaporation with hydrophilic membranes was investigated for the separation of methanol from the methyl esters. Pervaporation experiments with methanol/methyl acetate mixtures, the bottleneck in the system, were performed with the commercial membrane Pervap^{TM} 4155-70. With the ternary mixture, the viability of separating methanol from both methyl esters was confirmed.

The implementation of alternative separation tools shows a high potential for raising the product yield and for reducing energy demand. Simplification of process concepts satisfies the specific needs of the biobased industry.

529812 CFD Modeling of Turbulent Reactive Flows with Bourne Chemistry Comparisons

Abstract:

Reactive Computational Fluid Dynamics (CFD) simulation can be a powerful tool in designing industrial mixers. Many industrially-relevant reactive flows involve multiple parallel, series and competing reactions of different time scales. Slow reactions (Damköhler number<<1) are limited by the kinetics rate. Faster reactions (Damköhler number >>1) are limited by the mixing rate of the reactants. An accurate representation of the key competitive reactions as well as the turbulence effect, in a CFD model or during experiments using a model chemistry, is of key importance for mixer/reactor design.

This presentation aims to both highlight how these issues have been addressed from an industrial standpoint as well as expose the technological gaps that exist in the current experimental and modeling tools. The study case consists of a converging jet in cross-flow and includes a kinetic scheme with series-parallel reactions. A laboratory representation of a portion of the industrial mixer/reactor was constructed and reactive (the fourth Bourne chemistry: 2, 2 DMP and caustic neutralization with acid) flow experiments were conducted. The concentration of the NaOH, which reacts with the limiting reagent (acid), was varied. Several turbulent mixing-limited reactions models (Eddy Breakup Model, Eddy Contact Model and Multiple-Time-Scale Model) were compared with the reactive flow data (2,2 DMP conversion to acetone and methanol).

This work has revealed a number of gaps in the current modeling and experimental approaches. For example, the mixing-limited reagent is often diluted with a solvent and reacted with an excess co-reagent. These engineering controls are not accurately depicted by some of the mixing-limited reaction models. Other key issues include the resolution the turbulent flow field and the need for models to simulate the mixing and reaction of intermediates. This presentation will explain these gaps and invite the academic community to provide solutions.

534878 Validation of the Coalescence-Dispersion Model for Complex Chemical Reactions

Abstract:

ABSTRACT

"Validation of the Coalescence-Dispersion Model for Complex Chemical Reactions"

Gary K. Patterson, Professor Emeritus

Missouri University of Science and Technology

Chemical and Biological Engineering

Rolla, Missouri 65401

The integration of the coalescence-dispersion (c-d) model of mixing into a CFD simulation code has been extensively tested. The c-d model simulates mixing by assuming that the fluids may be represented by mobile fluid points which carry the mass, concentrations and other properties of the fluids. Upon coalescence of two randomly chosen fluid points, their properties are averaged (concentrations, enthalpies, etc.). Upon dispersion (separation) the two fluid points have identical properties. In a c-d simulation of a chemical reaction, the reactions occur for the length of the time increment between flow events as if the fluid points are batch reactors. Use of the c-d model avoids the need for a closure model for the mixing in the presence of chemical reactions. Such closures have been successful only for single reactions and for two competing reactions.

Integration of the c-d model into a CFD simulation requires that the mobile fluid points move, mix and disperse in a way dictated by the CFD simulation. Each volume segment of the CFD simulation is assigned a number of fluid points which move from segment to segment as dictated by the CFD simulation. This is accomplished by a subroutine linked to the CFD code that accesses the velocity and mixing rate predictions of the CFD. A critical aspect of the c-d model is in the most effective choices of the time increment and number of fluid points in each CFD volume segment. This is dependent on the number of c-d events during each time increment and flow volumes from segment to segment. Of course, the accuracy of the CFD simulation is critical for the accuracy of the c-d simulation.

Validations of c-d simulations have been done by comparing results with well-established experimental data for single and multiple reactions in various types and sizes of chemical reactors. Examples of validation data used are experiments by Paul and Treybal, Penney and coworkers, Middleton and coworkers, Bourne and coworkers and by the author and students. The validations show that the c-d simulation method can be a valuable addition to experimental research on multiple competitive reactions where there is interest in yields of chosen products.

537344 A Direct Numerical Simulation of a Marginally Turbulent Stirred VesselÂ Equipped withÂ a Rushton Turbine

Abstract:

New results will be presented from a Direct Numerical Simulation (DNS) of the just turbulent flow field, at Re=7300, in a stirred vessel provided with baffles and driven by a Rushton turbine. The simulations have been carried out by means of a Lattice Boltzmann solver, written in C, on a lattice comprising some 30 billion nodes, carrried out at the Irish Centre for High-End Computing ICHEC in Dublin.

The results of this simulation are compared with experimental data from the literature and with the results of the less refined DNSs reported earlier. The main variables of interest are the turbulent kinetic energy and its dissipation rate, phase averaged as well as phase resolved, which can be calculated from the outcome of such a DNS at any position in the vessel. Also turbulent spectra can be obtained in this way and some spectra will be presented for some positions in the vessel.

535761 Blending of Miscible Fluids

Abstract:

The blending process of binary miscible fluids is governed by both fluid advection and fluid diffusion. In low Reynolds number systems, where flow is slow and often one-dimensional, blending is diffusion-limited, and the effects of advection on blending are small. In high Reynolds number systems, where the fluids are fully turbulent, miscible fluid blending is governed by advection and the effects of diffusion are minimal. Between these two extremes, the blending behavior is a more complicated hybrid of both transport processes. This complexly is exacerbated in systems with strong difference between the two fluid densities and viscosity.

Traditional computational approaches for predicting the blending of miscible fluids are successful only in the limit of very low or very high Reynolds numbers. The limitation is motivated by the spurious effects of numerical diffusion, which tend to overstate blending. In this work, we show how DNS and LES simulations can be combined with high-resolution flux-limiting schemes to predict the blending of miscible fluids across a range of Reynolds and Schmidt numbers. After discussing strategies for controlling numerical diffusion, we apply this approach to model miscible fluid transport in both agitated systems and continuous flow systems.

538613 Deconstruction of Mixing Processes Using Computational Fluid Dynamics and *Z*-Transform

Abstract:

Computational fluid dynamics (CFD) is used in studying the performance of mixing processes to complement experimental work by making accessible to the investigator measurements that are difficult to obtain physically. Motivated by the desire to enhance the modeling toolset, we have in recent years demonstrated the use of CFD with *z*-transform, the discrete-time counterpart of the Laplace transform, to provide a powerful approach for simulating and analyzing the dynamics of mixing processes. Examples that we have illustrated include the open-loop response of a lumped-parameter mixing process, the open-loop response of a distributed-parameter mixing process, the closed-loop response of an unstable mixing process, and the general characterization of mixing processes [AIChE Annual Meeting Papers 2014-354913, 2015-404589, 2016-449915, and 2017-486416].

By the conventional approach, continuous-flow mixing processes are typically modeled as ideal continuous stirred-tank reactors (CSTR) or plug-flow reactors (PFR), or combinations thereof. Real mixing processes, however, do not conform perfectly to the CSTR and PFR paradigm. In order to optimize real mixing processes or to troubleshoot problematic ones, it is desirable to transcend beyond the CSTR and PFR blackbox approach and describe the dynamic behavior of the nonideal mixing processes more accurately. In this presentation we extend our previous work on combining CFD and *z*-transform and demonstrate how one can deconstruct nonideal mixing processes into networks of smaller and simpler mixing and transport processes. The result allows us to identify and construct practical mathematical models that capture localized phenomena such as compartmentalization and short-circuiting in nonideal mixing processes.

541565 25 by 25: Chemical Engineering in the Next 25 Years

Abstract:

In 2008, the centennial year for AIChE, we asked 25 chemical engineers for their visions of the profession from around the US and the world: Nine faculty, eight from industry, and eight post-docs and graduate students. We posed four questions related to each individual's industrial sector or area of research, asking for extrapolation, new impacts on existing sectors, new sectors, and an optional, more open-ended comment on the future of the profession as a whole.

In this 110th year of AIChE, we have revisited these questions with many of the same people. Ten years has seen many developments in our profession, in society, and in our thought-leaders' experiences. Here we report on their perspectives, comparing them to the 2008 visions.

541542 The Future of Chemical Engineering Itself

Abstract:

Chemical engineeringâ€™s future is bright because it is vital to so much of society and industry -- and its future is complicated by its rapidly expanding universe of activities. It must represent depth in core skills centered on the chemical sciences including biology and materials. At the same time, it must be transdisciplinary; it must exemplify the need and value of reaching across disciplines in borderless ways.

One way is to create and disseminate a compelling vision and strategy for chemical engineering, research, and education for the future. AIChE and the National Academies are working together to that end. In 1988, the National Academies published "Frontiers in Chemical Engineering: Research Needs and Opportunities.â€ Its committee of authors outlined a wide-ranging roadmap to turn promising research opportunities into reality, while calling for researchers to embrace new frontiers.

Since 1988, that has happened -- along with many other transformative changes like the Web, fracking, personalized medicine, and social media. A new nationwide effort, aided by international contributions, will address both the identity of chemical engineering and its future directions. Its goal is to ensure that the people in chemical engineering are primed to meet society's needs and rapid changes.

544708 Accelerating Innovation through Academic-Industrial Partnerships

Abstract:

In 2011, The Dow Chemical Company launched the University Partnership Initiative (UPI) to sponsor collaborative research at leading U.S. Universities. This 10-year commitment aims to address fundamental questions in industrial chemistry and chemical engineering with alignment to Dowâ€™s next-generation process technology and product development efforts. To date, this portfolio of projects has generated an impressive array of intellectual property, publications, and numerous examples of the collaborative developments implemented at Dow. This presentation will introduce UPI program and highlight several contributions from our collaborators and leaders in the chemical engineering profession.

540559 Maximizing Uptime, Efficiency, and Safety of Industrial Operations through Early Risk Detection

Abstract:

Achieving zero incidents while maximizing process uptime and efficiency is a key vision for operating companies. An important building block towards this vision is establishment of a proactive risk mitigation culture, supported by effective systems and a strong management. Studies have shown that most of the incidents and unexpected process failures can be avoided if the operating teams get timely information about developing risks and take preventive actions early on. While advances have been made in process monitoring and automation in the past decades, there still remains significant technological and behavioral gaps that prevent plant operations from proactively mitigating process risks.

A modern industrial plant monitors thousands of parameters, generating upwards of 50-100 million data points every day. Thanks to recent breakthroughs in machine learning and artificial intelligence approaches, today there are autonomous systems that can sift through this data and point out meaningful and timely insight. This can help operating teams ascertain process issues that are hidden in the data, long before process variables reach critical levels.

In this presentation, a new autonomous system, Dynamic Risk Analyzer™ (DRA), will be introduced that points out early indicators of process issues using proprietary machine learning. The risk indicators enable operations team to respond and make required process changes, days and sometimes weeks ahead of any alarm, avoiding late stage (expensive) fixes. To realize these objectives fully, a proactive management workflow must also be in place. The presentation will discuss new initiatives and workflows developed by plant operations to implement a proactive culture. Real-life case studies from plants using DRA technology will be presented on how this culture change resulted in increased uptime, efficiency and safety of plant operations.

531309 Gaussian Processes for Hybridizing Analytical & Data-Driven Decision-Making

Abstract:

Surrogate models are widely appreciated in process systems engineering [1]. The typical setting focuses on expensive-to-evaluate, possibly uncertain functions. Examples include: modular process simulators [2], integrated gasification combined cycle processes [3], a carbon capture absorber [4], and many other applications, e.g. [5-8]. Resources are typically limited, so effective decision making requires data-efficient learning.

The data science and statistical machine learning communities typically focus on models learned solely from observed data. But chemical engineering applications may also require explicit, parametric models, e.g. modeling known process constraints, operations constraints, and cost objectives [9]. So, work has integrated semi-algebraic functions with those learned from data [10] or developed semi-physical modeling techniques [11, 12].

This presentation surveys the state-of-the-art in hybridizing analytical and data-driven decision making [13]. We consider three probabilistic modeling applications to these hybrid situations:

Design of experiments for model discrimination [14]. We bridge the gap between classical, analytical methods [15] and Monte Carlo-based approaches [16]. Classical methods may have difficulty managing non-analytical model functions and data-driven Monte Carlo approaches come at a high computational cost. We replace the original, parametric models with probabilistic, non-parametric Gaussian process surrogates learned from model evaluations. The surrogates are flexible regression tools that extend classical analytical results to non-analytical models, while providing us with model prediction confidence bounds and avoiding the computational complexity of Monte-Carlo approaches.

Multi-objective optimization [17, 18]. We make novel extensions to Bayesian multi-objective optimization in the case of one analytical objective function and one black-box, i.e. simulation-based, objective function. The resulting method has been applied to a bone neotissue application [19] and a more general test suite.

Scheduling plant operations under uncertainty. For processes with equipment degradation, we use Gaussian processes to approximate large-scale, mixed-integer optimization problems.

We close by offering a broad outlook on applying probabilistic surrogate models to chemical engineering. Statistical machine learning has recently attracted significant interest in process systems engineering [20]. Here we show that state-of-the-art research in Gaussian processes [21, 22] and probabilistic modeling more generally [23] can have a big impact on chemical engineering.

References

[1] Bhosekar A, Ierapetritou M. Advances in surrogate based modeling, feasibility analysis and and optimization: A review. *Comput Chem Eng*. 108: 250-267, **2018**.

[2] Caballero JA, Grossmann IE. An algorithm for the use of surrogate models in modular flowsheet optimization. *AIChE J*. 54(10):2633-2650, **2008**.

[3] Lang Y, Zitney SE, Biegler LT. Optimization of IGCC processes with reduced order CFD models. *Comput Chem Eng*. 35(9):1705-1717, **2011**.

[4] Cozad A, Sahinidis NV, Miller DC. Learning surrogate models for simulation-based optim-ization. *AIChE J*. 60(6):2211-2227, **2017**.

[5] Boukouvala F, Ierapetritou M. Surrogate-based optimization of expensive flowsheet modeling for continuous pharmaceutical manufacturing. *J Pharm Innov*. 8: 131, **2013**.

[6] Soepyan FB, Cremaschi S, Sarica C, and others. Estimation of percentiles using the Kriging method for uncertainty propagation. *Comput Chem Eng*. 93:143-59, **2016**.

[7] Shokry A, Ardakani MH, Escudero G, Graells M, Espuña A. Dynamic Kriging-based fault detection and diagnosis approach for nonlinear noisy dynamic processes. *Comput Chem Eng*. 106:758-76, **2017**.

[8] Tran AP, Georgakis C. On the estimation of high-dimensional surrogate models of steady-state of plant-wide processes characteristics. *Comput Chem Eng*. DOI 10.1016/j.comp-chemeng.2018.02.014, **2018**.

[9] Boukouvala F, Hasan MF, Floudas CA. Global optimization of general constrained grey-box models: new method and its application to constrained PDEs for pressure swing adsorption. *J Glob Optim*. 67(1-2): 3-42, **2017**.

[10] Boukouvala F, Floudas CA. ARGONAUT: AlgoRithms for Global Optimization of coNstrAined grey-box compUTational problems. *Optim Lett*. 11(5): 895-913, **2017**.

[11] Pearson RK, Pottmann M. Gray-box identification of block-oriented nonlinear models. *J Process Contr*. 10(4):301-315, **2000**.

[12] Cozad A, Sahinidis NV, Miller DC. A combined first-principles and data-driven approach to model building. *Comput Chem Eng*. 73: 116-27, **2015**.

[13] Boukouvala F, Misener R, Floudas CA. Global optimization advances in mixed-integer nonlinear programming, MINLP, and constrained derivative-free optimization, CDFO. *Eur J Oper Res*. 252: 701-727, **2016**.

[14] Olofsson S, Deisenroth MP, Misener R. Design of Experiments for Model Discrimination Hybridising Analytical and Data-Driven Approaches. *arXiv:1802.04170*. **2018**.

[15] Buzzi-Ferraris G, Forzatti P, Emig G, Hofmann H. Sequential experimental design for model discrimination in the case of multiple responses. *Chem Eng Sci*. 39(1): 81-85, **1984**.

[16] Vanlier J, Tiemann CA, Hilbers PAJ, van Riel NAW. Optimal experiment design for model selection in biochemical networks. *BMC Syst Biol*, 8(20), **2014**.

[17] Olofsson S, Mehrian M, Geris L, Calandra R, Deisenroth MP, Misener R. Bayesian Multi-Objective Optimisation of Neotissue Growth in a Perfusion Bioreactor Set-Up. *Computer Aided Chemical Engineering*. 40: 2155-2160, **2017**.

[18] Beykal B, Boukouvala F, Floudas CA, Pistikopoulos EN. Optimal design of energy systems using constrained grey-box multi-objective optimization. *Comput Chem Eng*. DOI 10.1016/j.compchemeng.2018.02.017, **2018**.

[19] Mehrian M, Guyot Y, Papantoniou I, Olofsson S, Sonnaert M, Misener R, Geris L. Maximizing neotissue growth

kinetics in a perfusion bioreactor: An *in silico* strategy using model reduction and Bayesian optimization. *Biotechnol Bioeng*. 115(3):617-29, **2018**.

[20] Lee JH, Shin J, Realff MJ. Machine learning: Overview of the recent progresses and implications for the process systems engineering field. *Comput Chem Eng*. DOI 10.1016/j.compchemeng.2017.10.008, **2017**.

[21] Damianou A, Lawrence N. Deep Gaussian processes. In *Artificial Intelligence and Statistics*. 207-215, **2013**.

[22] Deisenroth MP, Ng JW. Distributed Gaussian processes. In *32nd International Conference on International Conference on Machine Learning*. 37: 1481-1490, **2015**.

[23] Ghahramani Z. Bayesian non-parametrics and the probabilistic approach to modelling. *Phil Trans R Soc A*. 371(1984):20110553, **2013**.

543554 Gas Capture by Ionic liquids

Abstract:

With the rapid development of the global industrialization process, a large number of gases, such as CO_2, SO_2, NH_3 and H_2S, have been produced. Among of them, CO_2 and SO_2 have caused air pollution problems, resulting in a serious harm to the environment and human health. On the other hand, CO_2 and SO_2 are also important source for synthesizing chemicals. Recently, ionic liquids (ILs) have been regarded as one of the most promising alternative solvents for capturing CO_2 or SO_2 due to their special physicochemical properties, such as negligible vapor pressure and designable character. For example, it was found that N-butylpyridinium thiocyanate ([C_4Py][SCN]) can absorb 0.841 gSO_2/g IL at 293 K and 0.10 MPa[1].

In this work, the mechanism of CO_2/SO_2 absorption in different kinds of ILs were systemically studied by molecular simulations. The microscopic structure, interaction energy and dynamic properties of the mixtures of ILs and CO_2/SO_2 were investigated. The results of interaction energy and fractional free volume of ILs showed that unlike non-polar gas CO_2, the interaction energy of cation-anion and free volume do not play a dominant role in SO_2 absorption. The solvation of SO_2 in ILs is controlled by enthalpy. Both the cations and anions play an important role in SO_2 solubility. Based on the above studies, a serial of ILs were synthesized for capturing CO_2/SO_2. Moreover, the energy and economic analysis of whole process of CO_2 capture with ILs were simulated. It was found that the IL process saved about 15% regeneration heat energy consumption and 11.5% total capture cost compared to the MEA process.

Acknowledgment

This work was supported by the National Natural Scientific Fund of China 51674234.

Reference

1. J. Zeng, H. S. Gao, X. C. Zhang, H. F. Dong, X. P. Zhang, S. J. Zhang, *Chem. Eng. J*, 2014, 251, 248-256.

533203 Thermoeconomic Optimization of Reverse Brayton Cycle Based Cryocooler for HTS Power Transmission Cable

Abstract:

In the era of the global sustainable energy crisis, an efficient power transmission system is required to meet the increasing power demand. After the discovery of superconductivity in 1911, many materials with zero resistance at a temperature below 25 K, known as Low-Temperature Superconductor (LTS), were discovered and a possible alternative for efficient power transmission. However, the cost of these materials and cooling system to maintain them in the superconducting state was very high. Thus, it is practically not feasible to use such materials for the transmission system. In 1986, the discovery of High Temperature Superconductors (HTS), which can be maintained in the superconducting state using liquid nitrogen, made the practical and economically feasible for use HTS power cable. These cables have to be maintained in a temperature range of 65 to 80 K with a cooling load of 2 to 5 kW for 1 km long transmission cable. Reverse Brayton cycle based Cryocooler (RBC) is the most suitable cooling system because of a wide range of cooling capacity at different temperature level. However, the capital and operational costs of RBC is high compared to the cryogenic roadmap (Gouge, 2004). The current cost of refrigeration for HTS devices is five to ten times of cryogenic roadmap ($25 per watt of refrigeration at 65-80 K). Hence, a thermoeconomic optimisation of RBC has been carried out to enhance the performance of the system from the thermodynamic and economic point of view.

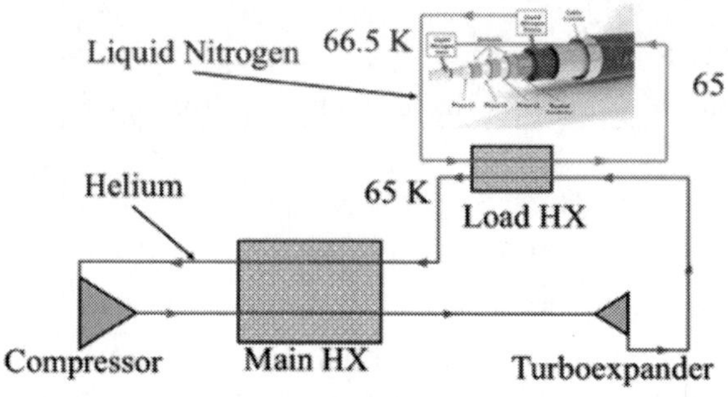

Fig. 1 Reverse Brayton Cycle based cryocooler for HTS power transmission cable.

In the present study, a reverse Brayton cryocooler consists of a compressor, main heat exchanger, turboexpander and load heat exchange as shown in Fig.1, has been investigated and optimised for 10 kW cooling load at 65 K. A commercial process simulator Aspen HYSYS V8.6 has been used for the study. Helium, more efficient than neon (Dhillon, 2017) has been used as working fluid for the cryocooler. The thermodynamic and economic models have been developed for analysing of the cryocooler. It is observed that the exergy efficiency of the system has maxima at 1.9 pressure ratio whereas cost per hour has manima at 2.3 pressure ratio as shown in Fig. 2. The inlet pressure and pressure ratio of the compressor have been varied to optimise the system process parameter. The effect of design parameters of equipment on the thermoeconomic performance of the system has been studied. The results obtained for RBC are presented in the present work and may be helpful for engineers and scientist to reduce the cost of RBC cryocooler without compromising the performance of the system.

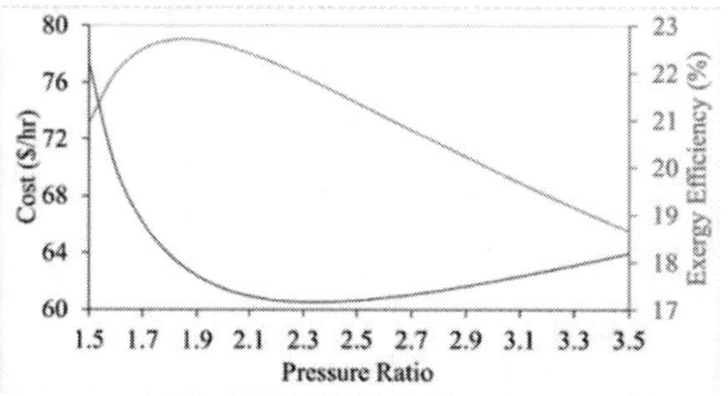

Fig. 2 Variation of cost per hour and exergy efficiency of RBC with different pressure ratios

1. Gouge, M. Cryogenic cooling technology for HTS electric machinery, 2004. IEEE Power Engineering Society General Meeting, 1-3, DOI: 10.1109/PES.2004.1373239.

2. Dhillon, A. K., Dutta R. and Ghosh P. Exergetic analysis of reverse Brayton cryocooler for different cooling loads at 65 k for HTS cables, 2017. 14th Cryogenics 2017 IIR International Conference, DOI: 10.18462/iir.cryo.2017.027

530055 Technology Readiness Levels As a Framework for Techno-Economic Assessment

Abstract:

Techno-economic assessment (TEA) is an important basis for decisions about research, development and deployment (RD&D). Methods applied in TEA depend heavily on the availability of data which is closely linked to the development progress. Technology readiness analysis and especially its rating scheme, technology readiness levels (TRL), are popular concepts for rating the maturity of RD&D according to available data. TRLs can therefore serve as a framework that facilitates the choice of methods along RD&D by sorting them according to their required input data.

TRL rating is widely adopted in the evaluation of RD&D projects by government (*e.g.* policy makers) [1,2], industry and academia. However, as the TRL concept was developed for space exploration [3], a domain that integrates a variety of disciplines from mechanical, electrical and chemical engineering to aviation and medicine, TRL guidelines and criteria remain unspecific to technologies, which can lead to misinterpretation and overly optimistic evaluation. A variety of TRL definitions evolved over the years, for example for biomanufacturing [4] or for the defense sector [5], defining specific TRL criteria for their respective domains. The need for a specification of TRL analysis for the chemical industry was identified in a report of the European association of research and technology organizations (EARTO) [6].

In this contribution, we summarize existing TRL concepts and propose adaptions and extensions to criteria for TRL rating in the chemical industry: Firstly, TRLs are characterized by the description of the project phase, its tangible deliverables and other qualitative criteria. Secondly, for a more objective and comprehensible TRL, specific indicators representing known information of various aspects of chemical engineering such as reaction engineering or process development are proposed. In addition, quantitative indicators such as the production capacity are often used and therefore discussed for different types of processes in this contribution.

Furthermore, we outline how methods of (capital) cost estimation and profitability calculation can be sorted by TRL in order to give guidance for sound TEA.

We gratefully acknowledge the financial support by the European Institute of Innovation and Technology (EIT) and the German Federal Ministry of Education and Research (BMBF).

[1] European Commission. EN HORIZON 2020 WORK PROGRAMME 2016 â€" 2017, General Annexes (European Commission Decision C (2017) 2468 of 24 April 2017), Annex G, Technology Readiness Levels (TRL); 2017.

[2] BMBF, Definition des Technologischen Reifegrades [â€¦] accessed on 08.02.2017 from https://www.ptj.de/lw_resource/datapool/_items/item_5502/definition_des_technologischen_reifegrades.pdf

[3] S. R. Sadin, F. Povinelli, R. Rosen, contribution at 39thIAF Congress, Bangalore, India, Oct. 8-15, 1988

[4] NSF Engineering Research Center for Biorenewable Chemicals (CBiRC). Technology Readiness Levels https://www.cbirc.iastate.edu/industry/technology-readiness-levels/ (accessed Oct 31, 2017).

[5] U.S. DEPARTMENT OF DEFENCE. Technology Readiness Levels in the Department of Defense (DoD) https://www.army.mil/e2/c/downloads/404585.pdf (accessed Oct 31, 2017)

[6] EARTO, The TRL Scale as a Research & Innovation Policy Tool, EARTO Recommendations, 2014

536017 Techno-Economic Modeling and Optimization of Catalytic Reactive Distillation for Bio-Oil Upgradation

Abstract:

The most significant benefit of pyrolysis bio-fuel is its availability from virtually every waste biological materials, which provides a double edge in sustainable waste management in addition to meeting our energy requirements. However, unfavorable characteristics of crude bio-oils received from pyrolysis process compared to petroleum fuels means that it must undergo upgradation before using in combustion engines. Reactive distillation (RD) is emerging as a powerful technology for these chemical reactive separation processes. RD, which has tremendous advantages over traditional reaction and separation unit operations including handling of unfavorable chemical and phase interactions, enhancing conversion and selectivity, circumventing azeotropic mixtures and reducing capital and energy costs, has been exploited for the reactive separation of various equilibrium controlled chemical reaction systems. Pyrolysis bio-oil upgradation is usually carried out by transesterification and esterification reactions with an alcohol and RD is suited pefectly to these types of equilibrium controlled reactions.

Here, equilibrium RD process simulation and economic evaluation was carried out for the esterification of n-butanol with complex mixtures of several fatty acids, phenolics and water (represent pyrolysis crude bio-oils) using Aspen PLUS process simulator and Aspen Process Economic Analyzer, respectively at both atmospheric and high pressure (10 bar). Complete design and optimization was carried out using evolutionary techniques from simplest system (ordinary distillation) to the most rigorous one (complex RD) with introduction of progressive complexities. RD simulation was performed using UNIQUAC as base-property method in Aspen RADFRAC module by minimizing Gibbs free energy. Prior to distillation design, esterification reaction kinetics and binary/ternary interactions were analyzed using Aspen RGibbs reactor module and Aspen Property PLUS, respectively. Simple distillation column was designed and optimized to obtain the initial trial parameters to be used in RD simulation, which was initially optimized only for acetic acid-water (75% acetic acid-25% water) mixture before RD design for complex fatty acid mixtures based on techno-economic feasibility and optimization. The reaction of phenol with fatty acids was incorporated and phenol was found to react with formic acid entirely, which increases the reaction conversion 5-10% due to its favorable effects. The conversions for the esterification reactions were found to be 88-99% for various simulated bio-oils with phenol composition 5-20% for butanol:acid (B:A) ratio of 2.5, and water was almost completely separated from the ester products which should increase the fuel quality. Higher water percentage in bio-oil feed was found to reduce acid conversion and ester separation. A reflux ratio of 0.95, distillate to feed ratio 0.505 and total 18 stages (3 rectifying, 9 reactive and 4 stripping) with n-butanol and bio-oil feed stage of 4 and 13, respectively, was found to be optimum at 1 bar based on reaction conversion, ester separation, heat duties and capital and operating costs. Effects of B:A feeding ratio and water percent in bio-oil is also analyzed and B:A = 2.25 was found to be optimum for 25% water in bio-oil. These works and optimized parameters can serve as a design platform for pyrolysis bio-oil upgradation to transportation fuels.

Reference:

1. Khan, M.A., Adewuyi, Y.G. High pressure reactive distillation simulation and optimization for the esterification of pyrolysis bio-oil. Process Engineering Journal, 1 (2017) 73-85.
2. Khan, M.A., Adewuyi, Y.G. Simulation of reactive distillation for esterification of pyrolysis bio-oil. The Proceeding of AIChE Annual Meeting, 2013, San Francisco, California, USA.

536658 Superstructure Optimization of Membrane-Based Carbon Capture Systems

Abstract:

This work presents the optimal design, configuration and operating conditions of a membrane-based post-combustion CO_2 capture process using computationally identified mixed matrix membranes. A new mixed integer nonlinear model is presented, which includes discrete and continuous decisions to minimize the cost of electricity (COE). The simultaneous optimization of process configuration, membrane materials, and operating conditions ensures rigorous exploration of the entire feasible design space.

Membrane systems often consist of multiple stages of membranes and compressors with intercooling. Flue gas source plays a very important role, since the CO_2 concentration highly affects the final plant configuration. While a simple multi-stage configuration may be optimal for concentrated CO
integrated with the boiler is often more effective for coal fired power plants.[1] Given the importance of finding the right configuration, two approaches have been used to optimize these systems. The first approach uses simulation-based optimization that leverages commercial process simulators such as Aspen Custom Modeler or gPROMS to develop rigorous models[2,3,4]. Previous work has used the Framework for Optimization, Quantification of Uncertainty and Surrogates (FOQUS)[2] to optimize the design and operating conditions of specific process configurations. In these approaches, the number of potential configurations that can be analyzed is limited since each must be set up individually.[3,4]

The second approach focuses on superstructure-based mathematical optimization of multi-stage membrane systems, which often requires simplified models for the equipment.[5,6] The advantage of the second approach is that the configuration as well as the design and operating conditions may be optimized simultaneously, resulting in more cost-effective plants.

Our work leverages efforts at NETL for developing new materials for membrane-based post-combustion capture processes.[2,3,4,7] An advanced superstructure-based process optimization model is developed using rigorous mathematical models for compressors, pumps, mixers, and heat exchangers and reduced order models for liquefier columns and property calculations. The current work makes use of results from a recent study[4], which analyzed a large number of hypothetical mixed matrix membrane materials for carbon capture. This work addresses the aforementioned limitations by analyzing multiple potential advanced process configurations and the selection of the membrane material from a wide set of potential options while optimizing the plant operation.

[1] Merkel T. C., Lin H., Wei X., Baker R. (2010). Power plant post-combustion carbon dioxide capture: An opportunity for membranes. Journal of Membrane Science. 359, 126-139.

[2] Miller D. C., Agarwal D., Bhattacharyya D., Boverhof J., Cheah Y. W., Chen Y., Eslick J., Leek J., Ma J., Mahapatra P., Ng B., Sahinidis N. V., Tong C., Zitney S. E. (2016). Innovative computational tools and models for the design, optimization and control of carbon capture processes, Computer Aided Chemical Engineering, 38, 2391-2396.

[3] Morinelly J. E. M., David C., 2012, Post-Combustion Gas Permeation Carbon Capture System Models, AIChE Annual Meeting, Pittsburgh, PA, October 28-November 2, 2012.

[4] Budhathoki S., Ajayi O., Steckel J.A., Wilmer C.E. (2018). High-throughput computational prediction of the cost of carbon capture using mixed matrix membranes. In preparation.

[5] Hasan M. M., Baliban R. C., Elia J. A., Floudas C. A. (2012). Modeling, Simulation, and Optimization of Postcombustion CO_2 Capture for Variable Feed Concentration and Flow Rate. 1. Chemical Absorption and Membrane Processes. Industrial and Engineering Chemistry Research. 51, 15642-15664.

[6] Arias A. M., Mussati M. C., Mores P. L., Scenna N. J. (2016). Optimization of multi-stage membrane systems for CO_2 capture from flue gas. International Journal of Greenhouse Gas Control. 53, 371-390.

[7] Zamarripa M. A., Ajayi O., Matuszewski M., Miller D. C., 2017, Superstructure-based optimization of membrane-based carbon capture systems. AIChE Annual Meeting, Minneapolis, MN, October 28-November 2, 2017.

543630 Making Hydrogen from Water with a Protein Organized Electrode: Ultra-High Utilization of Noble Metal in Proton Exchange Membrane Electrolysis for Capital Cost Reduction

Abstract:

Renewable hydrogen could play a critical role in low-carbon energy systems and in fulfilling intermittent energy needs. Proton exchange membrane (PEM) electrolysis technology for hydrogen production is safe, simple, and highly efficient. PEM electrolysis has ability to be integrated in a variety of energy systems and infrastructures, such as transportation, renewable electricity, and biogas upgrading. It also has the ability to address seasonal-variation in long timescales. But, one of the major challenges remains the high cost of noble-metal electrodes in the membrane electrode assembly (MEA). Conventional catalysts like platinum and iridium are required to facilitate the hydrogen evolution reaction (HER) and the oxygen evolution reaction (OER) at electrodes for superior stability in long-term. The contribution of catalyst cost in the MEA becomes a more significant fraction of the capital cost when the production is scaled up.

To reduce the capital cost, we have developed protein-integrated electrodes which are designed to allow lower amounts of noble metal catalyst while maintaining high electrolysis performance. The purpose of our designed protein, elastin-like polypeptides (ELP), is to organize ionomer networks and create efficient pathways between the ionomer and catalyst sites, thereby significantly improving catalyst utilization. A preliminary cost analysis indicates that even at current prices for engineered proteins and using reasonable ratios of protein to catalyst particles (~6.5 metal:protein by mass), a 10X reduction in noble metal could be achieved for the same cost. Considering this technology is expected to lower loadings by more than 10X, and that the future cost to produce these proteins will decrease while the cost of noble metals will increase, the technology has great economic potential. For example, if the cost of protein manufacturing drops by a reasonable 25%, and a 20X reduction in metal loading is achieved, the total electrode cost would be 60% lower. ELP is suitable as electrode organizing molecules for its flexibility in design, modular format, and manufacturability. Our results using a quartz crystal microbalance and a transmission electron microscope show a robust binding of ELP to platinum catalyst and ionomers which affects the resulting ionomer-catalyst structures. This system provides an innovative, multidisciplinary solution for improving catalyst utilization in membrane electrode assemblies.

522132 Design Optimization and Economic Evaluation of LNG/NGL Integrated Processes Under Lean Feed Composition

Abstract:

The economic evaluation of the integrated liquefaction and natural gas liquids (NGL) recovery process for the lean feed is required to meet recent increasing demand for lean gas reservoirs. This paper performs economic analysis for the conventional co-production of liquefied natural gas (LNG) and NGL process compared to the proposed simplified integrated process with consideration of the LNG higher heating value (HHV) specification under the lean feed condition. The dual mixed refrigerant (DMR) cycle was selected for liquefaction part due to high efficiency reported by previous studies. The schemes were built by Aspen HYSYS commercial software and the object function was optimized with a genetic algorithm (GA). The results show that the simplified integrated process has economic advantages compared to the conventional integrated scheme under the lean feed composition, which shows different results in terms of previous studies conducted by normal or rich feed condition. Additionally, the capital cost can be reduced remarkably by simplifying NGL recovery section with almost no loss of liquefaction efficiency in the proposed simplified integrated process.

526738 Intensifying Natural Gas Upgrading: Integrated Reactor Concepts for Syngas Production

Abstract:

The recent surge in economically accessible natural gas reserves has revived interest in the development of novel, less capital intensive, more resource efficient, and cleaner processes for natural gas conversion via upgrading to valuable chemical feedstocks. This development, often summarized under the label "Process Intensification" (PI), is based on a conceptual approach to process engineering that aims to radically reduce the physical, energetic, and/or environmental footprint of existing processes without altering the basic underlying process chemistry.

This presentation will discuss examples of process intensification principles applied to natural gas conversion, drawing on work from our own laboratory and others. A particular focus will be on the use of multifunctional reactor concepts, ranging from heat-integrated reactors to chemical looping. I will argue that multifunctionality is at the heart of PI, marking a significant departure from the "unit operations" approach that has defined chemical engineering since its inception in the early 20th century. Systematic application of PI can drastically increase efficiency, process safety, and reduce the overall footprint of next-generation natural gas conversion processes.

527943 Process Concept for Isolation of Low Molecular Weight Carboxylic Acids from Dilute Aqueous Feed

Abstract:

Market launch of biobased bulk products, such as low molecular weight carboxylic acids, alcohols and ketones, very often suffers from poor economics. Low concentration of target constituents is a major obstacle. Process intensification is a promising approach for addressing the economic feasibility of biobased bulk products.

Liquid-liquid extraction with chemical reaction in emulsified state offers excellent technological features for application in this challenge. This concept, successfully applied in multiphase homogeneous catalysis, was investigated for acetic acid isolation by esterification from dilute aqueous effluents with feed concentration of less than 60 g L^{-1}. Conversion rates of up to 55%, compared to conversion of just 4% esterification without catalysis were achieved. For further optimization of this process concept, investigating parameter interaction was necessary to maximize conversion. Design of experiments (DoE) was applied to evaluate relevant process parameters temperature, stoichiometric ratio and catalyst load in batch experiments. Evaluation of design of experiments revealed a possible mass transfer limitation for low phase ratio. Based on the DoE results a model was fitted with high reproducibility and agreement with experimental data. Accuracy of conversion prediction with the model was experimentally verified and confirmed. Downstream processing via transesterification by reactive distillation was applied to complete the technology concept.

*nâ€'*octanol, diluted in the solvent *nâ€'*undecane, was used as shuttle reactant. *n*-octanol was chosen because of its negligible water solubility. The anionic surfactant 4â€'dodecylbenzenesulfonic acid catalyzed the esterification reaction. Emulsification during conversion provided large mass transfer area. Regeneration of the shuttle reactant *n*-octanol was performed by transesterification with short chain aliphatic alcohols (e.g. methanol) after emulsion splitting by centrifugation. The low molecular weight products methyl acetate/methanol can simply be separated from the reaction broth by single step distillation due to the huge boiling point difference. It has to be mentioned that the bottom stream of the distillation step may be recycled to the esterification without temperature adjustment. Reaction kinetics of esterification and transesterification for varying stoichiometric ratio and operation temperature level were subject of detailed analysis. For validation, the process concept was further tested for the low molecular weight carboxylic acids formic acid and propionic acid.

537987 Process Integration Using Block Superstructure

Abstract:

Process integration enables the design of integrated production systems ranging from individual processes to total sites to reduce fresh resource consumption or harmful emissions [1-4]. A classic process integration problem consists of sinks with demand for certain properties, and sources to satisfy the demands of these sinks. These properties exchanged within the system can be materials/contaminants in water networks [5], mass exchange networks [6], etc. The properties can also be exchanged as energy, e.g., heat in heat exchanger networks [7], power in work exchange networks [8]. Capturing these recurring features helps to construct a unified superstructure representation for different integration problems. Besides, there is a symbiotic relation among process integration, synthesis and intensification [9-11], which necessitates the advancement of process integration techniques. Recently, the building block superstructure has been proposed for systematic process intensification by dissecting unit operations into fundamental phenomena building blocks [9]. Later on, this approach is applied for general process synthesis problems without postulating case-specific process superstructures [10].

In this work, we extend the block representation to incorporate various process integration problems by considering their common features. Process units, including regenerators/interceptors, are embedded in blocks. The feed flow and product flow in each block represent sources and sinks, respectively. Adjacent blocks interact with each other through direct connecting streams equipped with heaters/coolers for adjusting the temperature. Besides, we also introduce 'jump flow' among all blocks to avoid more intermediate blocks connecting nonadjacent blocks while maintaining or increasing the number of process alternatives. The block superstructure size is dependent on the number of layers with mixing operations, process units, product streams and heat integration stages. We formulate the general process integration problem as a mixed-integer nonlinear optimization (MINLP) problem to minimize the total annual cost. With the proposed MINLP model, we investigate various process integration problems involving mass integration, heat integration, simultaneous mass and heat integration, and property integration. These case studies showed that the same reported optimal integration networks and/or better network designs can be generated using the proposed block representation approach.

References:

[1] Li J., Demirel S.E., Hasan M.M.F. Process Integration using Block Superstructure. *Industrial & Engineering Chemistry Research*, 2018, 57: 4377–4398.

[2] El-Halwagi, M. M. Pollution prevention through process integration. *Clean Products and Processes*, 1998, 1: 5–19.

[3] Friedler, F. Process integration, modelling and optimization for energy saving and pollution reduction. *Applied Thermal Engineering*, 2010, 30: 2270–2280.

[4] Li J., Demirel S.E., Hasan M.M.F. Fuel Gas Network Synthesis Using Block Superstructure. *Processes*. 2018, 6(3): 23.

networks. *AIChE journal*, 2011, 57: 434–457.

[6] El-Halwagi, M. M.; Manousiouthakis, V. Synthesis of mass exchange networks. *AIChE journal*, 1989, 35: 1233–1244.

[7] Yee, T. F., Grossmann, I. E. Simultaneous optimization models for heat integration II. Heat exchanger network synthesis. *Computers & Chemical Engineering*, 1990, 14: 1165–1184.

[8] Razib, M., Hasan, M. M. F., Karimi, I. A. Preliminary synthesis of work exchange networks. *Computers & Chemical Engineering*, 2012, 37: 262–277.

[9] Demirel, S. E., Li, J., Hasan, M. M. F. Systematic process intensification using building blocks. *Computers & Chemical Engineering*, 2017, 105: 2–38.

[10] Li, J., Demirel, S.E., Hasan, M.M.F. Process Synthesis using Block Superstructure with Automated Flowsheet Generation and Optimization. *AIChE Journal*, 2018, under review.

[11] Baldea, M. From process integration to process intensification. *Computers & Chemical Engineering*, 2015, 81:104–114.

535411 The Properties of Gases and Liquids: 2020

Abstract:

For decades, The Properties of Gases and Liquids (PGL) has been a standard reference for professionals in the field, advanced undergraduate, and graduate students. With 27,000 google scholar citations and counting, this book serves a similar role to "Consumer Reports" in objectively evaluating the range of estimation methods available in the literature. Nevertheless, this important resource has not been updated since 2001. The current authors have undertaken to provide this update, with careful attention to the many changes in technology and methods that have occurred over the years. Our mission is to survey the entire field of physical property estimation in order to promote the best approaches and objectively characterize the quality of the available data and methods, such that current applications are reliable and future progress of the field is advanced.

A lot has happened since the fifth edition of PGL was published. Broadly speaking, advanced computer methods have become much more feasible for practical engineering calculations. These include quantum mechanical (QM) methods like Moller-Plesset and quantum density functional theory (DFT), quantitative structure property relations (QSPR; both closed-form and machine-learning methods), and molecular simulations like Monte Carlo (MC) and molecular dynamics (MD) methods. Additionally, group contribution (GC) methods have become more sophisticated, with second and third order groups in many cases. Despite the increased sophistication, these approaches are not guaranteed to be more accurate than traditional methods, so rigorous evaluation of all available methods, both old and new, is essential for practical application. Doing so requires the specification of a common set of metrics against which the performance of all methods is compared. This set of metrics cannot simply be obtained from the original literature as the range of compounds, conditions, and comparative statistics varies from study to study. As in previous versions of PGL, care is taken in this edition to define an objective and comprehensive set of metrics for proper evaluation of existing methods so that proper recommendations may be made about predictive methods.

Modern chemical databases and computer automation make evaluation possible in a manner not available previously, but this increased capacity for objective comparisons is consistent with PGL editions of the past. For many years, the spirit of PGL has reflected the working hypothesis that predictive methods are most reliable when they combine the fundamental principles of physics as a foundation with an appropriate degree of empiricism to ensure accurate reflection of experimental measurements. In this sense, experimental measurement comprises the final arbiter of any engineering model. Therefore, the first order of business in evaluating physical property methods is the compilation of standard databases for comparing the methods on a consistent basis. Despite best intentions, it would not be feasible for these authors to code and test every method that has been published since the fifth edition. The advancing sophistication of predictive models has made it more difficult to reproduce and evaluate many methods. Therefore, we envision a two stage process for making comparisons. In the first stage, method developers will submit working executables of their method in accordance with working protocols that we define. These submissions will be evaluated against the databases and the top few methods will be selected for the sample calculation stage. In the second stage, sample calculations will be performed by the authors. Final recommendations will include the sample calculations in textbook form along with a possibly subjective consideration of how easily reproducible the calculations would be for our readers to perform.

In this presentation, we outline our expectations for database development and method comparisons in greater detail.

528118 Computer-Aided Polymer Design Using COSMO-RS

Abstract:

COSMO-RS[1] is a successful and accurate statistical thermodynamics method that has distinct advantages for computer-aided molecular design (CAMD)[2]: (1) it incorporates QM-derived electronic and structural information; (2) thermodynamic property estimation of fluids is extremely fast assuming that a unique surface charge distribution called a sigma profile is known for each species in solution; and (3) COSMO-RS as a technique is amenable to integration with common CAMD strategies[3,4]. However, CAMD using COSMO-RS is still a developing research area and has thus far typically focused on small molecule and solvent design. Addressing polymeric systems with COSMO-RS-based CAMD opens these powerful design techniques to diverse applications in membranes, electronics, aerospace materials, green chemistry, etc. Polymer design with COSMO-RS also benefits from the methodâ€™s accuracy in this domain as has recently been shown for a few properties[5].

We address the optimization of polymers in this work with two primary strategies. The first leverages the simplicity of calculating the sigma profiles of polymeric structures from monomeric sub-units[5]. This strategy naturally leads to a mixed-integer nonlinear programming (MINLP) formulation which allows for the design of a copolymer from a large number of monomeric candidates. As a second strategy, we project the high-dimensional search space of polymeric structures into a lower dimensional space of the sigma moments (the statistical moments of the sigma profiles). This allows for the use of state-of-the-art derivative-free optimization (DFO) algorithms to address the design problem in a meaningful lower-dimensional space, for which they have been shown to be effective at finding good solutions in CAMD problems[3]. Because this strategy designs structures at each iteration, we also discuss the possibility of using molecular dynamics (MD) simulations for each designed system to improve the accuracy of the property estimates. We apply the methods to a few case studies drawn from the polymer literature.

1. Klamt A, Jonas V, Bürger T, Lohrenz JC. Refinement and parametrization of COSMO-RS. The Journal of Physical Chemistry A. **1998** Jun 25;102(26):5074-85.
2. Austin ND, Sahinidis NV, Trahan DW. Computer-aided molecular design: an introduction and review of tools, applications, and solution techniques. Chemical Engineering Research and Design. **2016** Dec 1;116:2-6.
3. Austin ND, Sahinidis NV, Trahan DW. A COSMO-based approach to computer-aided mixture design. Chemical Engineering Science. **2017** Feb 23;159:93-105.
4. Scheffczyk J, Fleitmann L, Schwarz A, Lampe M, Bardow A, Leonhard K. COSMO-CAMD: A framework for optimization-based computer-aided molecular design using COSMO-RS. Chemical Engineering Science. **2017** Feb 23;159:84-92.
5. Loschen C, Klamt A. Prediction of solubilities and partition coefficients in polymers using COSMO-RS. Industrial & Engineering Chemistry Research. **2014** Jul 7;53(28):11478-87.

534840 Design of Electrosprayed Polysaccharides Nano/Microparticles for Drug and Vaccine Delivery

Abstract:

Polysaccharides are natural polymers that are attractive for a variety of applications because of their renewability and biocompatability. Particles prepared from polysaccharides are of interest as carriers for controlled release in food, cosmetics and pharmaceuticals, and as platforms for biosensing in environmental and biomedical fields. Electro-hydrodynamic atomization or electrospraying is an attractive process, in terms of controllability and scalability, for synthesizing such particles. The objective of our study was to investigate the effects of individual factors in the preparation of polysaccharide particles by the electrospray process on the particle size and shape using factorial design and response surface methodology. Two types of polysaccharides, namely the anionic alginate and the cationic chitosan, were investigated. The response surface method (RSM) experiments were conducted for sodium alginate based on the results of the one-half fractional-factorial design with five factors (flowrate, voltage, polymer concentration, collection distance, and surface tension of the collection medium) and two levels (low, high). The microgel particle size of sodium alginate was found to be primarily dependent on the polymer concentration, flowrate, and applied voltage. Formation of non-spherical particles was also observed for certain molecular weights of the polymer. Similar experiments were performed with for chitosan, with molecular weight of chitosan, polymer concentration, acetic acid concentration, solution flowrate, voltage, nozzle diameter and collection distance as factors. Toward the development of a process for protein-encapsulation, electrospraying of proteinâ€"chitosan polyelectrolyte complexes were studied. Bovine serum albumin was used as a representative low-isoelectric-point protein. Split-plot RSM with seven factors was conducted to understand the effects on particle size and protein encapsulation efficiency.

References:

Myrick, JM. Stimulus-Responsive Polysaccharide Microgel Particles Prepared By Temperature-Induced Aggregation and Electrospray Techniques. Ph.D. Dissertation, Clarkson University, May 2017

Myrick JM, Vendra VK, Krishnan S. Self-Assembled Polysaccharide Nanostructures for Controlled-Release Applications. *Nanotechnol. Rev.* 2014, 3, 319â€"346, DOI: https://doi.org/10.1515/ntrev-2012-0050

Sexton FA, Krishnan S, Vendra VK. Sustained Release of Nutrients In Vivo. Patent No. US 8563066 B2 (2013)

538371 QSAR Study of Combretastatin-like Chalcones As Cancer Cell Growth Inhibitors Using Linear and Non-Linear Machine Learning Approaches

Abstract:

Combretastatin-like chalcones are known to be effective inhibitors for the mitotic process that tends to help the growth of cancer cells. Chalcones interact with δ±-tubulins, and inhibit the polymerization process that ends up increasing tumor size [1]. Based on the works of Ducki et al [2], 87 chalcones were selected. It is important to mention that an attempt to model inhibitory concentration (pIC_{50}) was made by Lipinski et al [3]. Although they have achieved models with acceptable fitness, there is still room for improvements. It can be noticed in their work that, although they used a variety of descriptor classes, their use of model development algorithms is lacking in light of recent advances.

This project aims at modifying, improving, and implementing machine learning and model development algorithms such as Artificial Bee Colony, Multi-gene Genetic Algorithm, Particle Swarm Optimization algorithms in combination with linear, non-linear, and LASSO regression algorithms. The goal is to develop models with better fitness using descriptors that are comparatively easier to calculate (e.g. connectivity descriptors, 2D descriptors). Avogadro open access software was used to develop and optimize the molecular structures, and Dragon 6.0 software was used to generate descriptors for the given compounds. One-fifth of the samples were used for external validation of model. Model superiority was decided based on values of R^2, Q^2, MSE, and MAE values of the models.

Selected References:

[1] Sharma et al, A review of mechanisms of antitumor activity of chalcones, 2016, Anticancer Agents Med Chem, 16, 200-211.

[2] Ducki et al, Combretastatin-like chalcoles as inhibitors of microtubule polymerization. Part 1: synthesis and biological evaluation of anitvascular activity, 2009, Bioorg Med Chem, 17, 7698-7710.

[3] Lippinski et al, A molecular modeling study of Combretastatin-like chalcones as anticancer agents using PLS, ANN and consensus models, 2018, Struct Chem (2018), https://doi.org/10.1007/s11224-017-1072-2

524501 Integration of Heuristic Knowledge in a Skin-Care Emulsion Design

Abstract:

The design of optimal mixtures is considered an important challenge in many industrial sectors (Cheng et al., 2009; Gani, 2004; Lee et al., 2014), mainly due to the need for a limited searching space for the most appropriate components, and their level of use within the final formulation. In recent years, many modelling approaches have been proposed to systematize and accelerate this screening process. Such approaches usually integrate rule-based elements and/or experiments in a combinatorial modelling problem with highly nonlinear expressions relating composition and product properties (Smith and Ierapepritou, 2010). Besides, in the industrial domain of cosmetic formulations, there is a considerable amount of available heuristic knowledge, regarding qualitative function of ingredients, as well as their impact on sensorial attributes. In a previous study (Arrieta-Escobar et al., 2017), a CAMD methodology (Achenie et al., 2002) integrated with heuristics was proposed for the case of hair care emulsions using mixed-integer optimization with logical constraints (Williams, 2013).

Here we illustrate the CAMD methodology integrated with heuristics and its use in the cosmetic emulsions design, using a body lotion as an example. Initially, the general structure of the methodology is presented; then, the heuristics regarding formulation and manufacture, the selected performance parameters for this kind of products, and the details of the experimental validation are described. The results show that the incorporation of heuristics into a systematic CAMD methodology using integer programming with logical constraints enables to obtain a set of feasible alternatives, which are optimised to meet the specified requirements of performance with a minimal cost. The obtained emulsion design was validated by developing and manufacturing a set of body lotion samples. The rheological, textural and microstructural properties of the samples, along with their consumer assessment, show that these alternatives met the expected properties. This methodology could be used as a rapid preliminary design tool for inexperienced cosmetic scientists and designers. Also it can help to reduce the time and resources spent in cosmetic emulsions design, comparing with the traditional trial-and-error methods.

References

Achenie, L.E.K., Gani, R., Venkatasubramanian, V. (Eds.), 2002. Computer Aided Molecular Design: Theory and Practice. Elsevier, Amsterdam.

Arrieta-Escobar, J.A., Bernardo, F.P., Orjuela, A., Camargo, M., Morel, L., 2017. An Integrated Methodology for Emulsified Cosmetic Product Formulation Using Integer Programming with Logical Constraints, in: Espuña, A., Graells, M., Puigjaner, L. (Eds.), Computer Aided Chemical Engineering. Elsevier B.V., Barcelona, Spain, pp. 985–990. https://doi.org/10.1016/B978-0-444-63965-3.50166-5

Cheng, Y.S., Lam, K.W., Ng, K.M., Ko, R.K.M., Wibowo, C., 2009. An integrative approach to product development—A skin-care cream. Comput. Chem. Eng. 33, 1097–1113. https://doi.org/10.1016/j.compchemeng.2008.10.010

Gani, R., 2004. Chemical product design: challenges and opportunities. Comput. Chem. Eng. 28, 2441–2457. https://doi.org/10.1016/j.compchemeng.2004.08.010

Lee, C.K.H., Choy, K.L., Chan, Y.N., 2014. A knowledge-based ingredient formulation system for chemical product development in the personal care industry. Comput. Chem. Eng. 65, 40–53. https://doi.org/10.1016/j.compchemeng.2014.03.004

Smith, B. V., Ierapepritou, M.G., 2010. Integrative chemical product design strategies: Reflecting industry trends and challenges. Comput. Chem. Eng. 34, 857–865. https://doi.org/10.1016/j.compchemeng.2010.02.039

Williams, H.P., 2013. Model building in mathematical programming, 5th ed. John Wiley & Sons, Inc., Chennai.

536301 Investigating Fluid-Particle Interactions in Expanded Beds Using CFD-DEM

Abstract:

The purification of high value products such proteins produced via fermentation typically involves a cascade of unit operations such as membrane filtration and chromatography columns. Expanded bed adsorption (EBA) creates an opportunity for process intensification by combining cell removal with a product capture and initial purification step. By mildly fluidizing a bed of adsorbent particles, the inter-particle void space is increased, which allows cells and other particulate matter to move through the bed in a relatively unhindered manner. The bed is stabilized through the use of a particle size and density distribution, which cause bigger/heavier particles to remain in the bottom while smaller/lighter particles reside higher up in the column. This results in a gradient in particle size, particle density and particle volume fraction along the height of the bed which reduces particle movement and thereby reducing unwanted back mixing in both phases.

To successfully scale up this technology to production scale, a more fundamental understanding of the mechanisms behind fluidization in EBA columns is required. Research into expanded beds using experimental techniques yields different conclusions based on whether the solid or liquid phase is examined (1, 2). Given the nature of expanded bed systems (dense, opaque multiphase flows), non-intrusive measurements of the fluid-particle interaction at the particle scale are highly challenging.

As an alternative to experimental measurements, computational fluid dynamics (CFD) can be combined with a discrete element model (DEM) for the particle phase. This work will present the results of CFD-DEM simulations on lab scale columns and compare them with experimental data. It will be demonstrated how this approach can be used to link the mesoscopic behaviour of the particles to macroscopic phenomena such as bed expansion, fluid phase dispersion and velocity. Based on the insights obtained from the simulations, resin properties such as density and size distribution can optimized to ensure bed stability and separation performance. The overall goal is to understand, optimize and scale EBA technology which then can be used for various business applications, like bio-pharmaceuticals, natural products and enzymes. References

1. Souquet J, Liu HY, Fan XF, Theodosiou E, Thomas O. Exposing expanded beds to PEPT. Journal of Biotechnology. 2007;131(2):S191-S.
2. Lin DQ, Shi W, Tong HF, van de Sandt EJ, Boer P, Ferreira GN, et al. Evaluation and characterization of axial distribution in expanded bed: II. Liquid mixing and local effective axial dispersion. J Chromatogr A. 2015;1393:65-72.

COMPARISON OF FIRST PRINCIPLES AND PARAMETRIC MODELS FOR THE DESIGN AND OPERATION OF A SPRAY DRYER FOR WHEY PROTEIN PRODUCTION

ANIBAL JOSÉ BARRIOS QUANT* 1, JOHN TELOTTE 2, JOSE ANDRES PEREZ MENDOZA 1.

1, Chemical engineering department, Universidad del Atlántico, Km 7, Puerto Colombia, Atlántico, Colombia

2, Department of Chemical and Biomedical Engineering, FAMU-FSU College of Engineering, Florida State University, 2525 Pottsdamer St, Tallahassee, FL, USA

*Corresponding author:abarriosquant@fsu.edu; Tel: +57 3053058185

Abstract

In this work we analyze four models for spray dryer design for a whey protein process. The models include both first principles approaches and empirical correlations. For a large set of given set of design specifications (inlet and outlet moisture contents, inlet air conditions and residence time) the physical size of the spray drying system was determined from all correlations and then the designs were compared to both industry heuristics for size ratios and actual dryer performance (where possible) to assess the validity of the models. Specific consideration was given to the correlation between the systems that met design heuristics and those that produced usable product. An economic analysis was developed to compare the prices that could be obtained designing a spray dryer based on the Masters restrictions. Finally a simulation on ANSYS CFD represented the flowfield of the spray dryer was developed.

KEYWORDS: Spray Dryer, design, sizing, dimensional analysis, critical variables, whey protein.

INTRODUCTION

Whey proteins with improved flavour and increased functionality are obtained with new technologies, as spray drying. While choosing a particular whey product, it is essential to match its function to the characteristics we want to achieve. For example, high protein concentrates or isolates are usually produced in spray dryers due with this process is possible to guarantee high quality powder [1,2].

Spray-drying is a rapid, continuous, cost-effective, reproducible and scalable process for the production of dry powders from a fluid material by atomization through an atomizer into a hot drying gas medium, usually air [3]. The design of spray dryers has remained largely empirical; it is still mainly based on the extensive experience and the vast body of operating data which the manufacturers have acquired over the years. To design spray dryers usually designers develop assumptions based on variables that allow an easier design process for spray dryers [4]. The focus of this study is to compare four different design models and their responses over common variables adjusted to evaluate the sizing of a spray dryer for whey protein production.

COMPARATIVE MODELS

The numerical equations who represented a phenomenon is better called design model, and is commonly used as a representation of spray dryers as well as digital simulations of spray drying. In this study the method was applicable to represent the particle movement inside the spray chamber and the dimensional structure of the equipment.

The dimensional structure of the spray dryer was explained by the models of Mujumdar [7], Suryanarayana [8], while the behavior of the drops inside the spray and the atomization process were showed by the models of Gluckert [9] and Masters [10].

The comparative analysis of these models was focused on study each models and evaluate their responses over the same operational conditions. The equations below show the structure of the models represented only by the diameter, length and volume. The complete structure of each model can be found in the supplementary information.

Mujumdar Model [7]

$$D = \sqrt{4\,G / 3600\,\pi\,j\,u}$$

$$L = 4V / \pi D^2 \qquad (1)$$

$$V = t\,F_2 / H\;\rho_{whey}$$

Suryanarayana Model [8]

$$D_{Spray} = \sqrt{4\,A_c / \pi}$$
$$L = 0.6\,D_{spray} \qquad (2)$$
$$V = G_s\,V_{avg}\,t_d$$

Gluckert Model [9]

$$D = 2\sqrt{(t_T\,2.24\,V_{res}\,(b\,(D_r / 2)^{12}} + (D_r / 4)$$
$$L = 1.5\,D_{spray} \qquad (3)$$
$$V_t = G_s\,V_{avg}\,t_d$$

Masters Model [10]

$$Dvs = 14000\,Q\,10.24 / n\,h\,0.12\,N\,d\,0.6$$
$$L = 1.5\,D_{spray} \qquad (4)$$
$$V = 1/4\;\pi\,D_i\;2\,h_{spray}$$

The comparison of the four models was evaluated by the variation of key variables of the model in range adjusted by the study based on avoidance of whey protein denaturation. The key variables were selected according to their influence over the four models. The values assigned to the key variables were reported in the literature and are shown in Table 1.

Table 1. Variables and values

Independent Variables (Factors)	Values	
X1= Humidity content (kg/kg db)	0.60	0.90
X2= Inlet air temperature (°C)	85	180
X3= Outlet air temperature (°C)	80	120
X4= Outlet whey humidity (kg/ kg dry)	0.05	0.10
X5= Air flow (m3/h)	123	288

Key variables

The key variables were organized randomly and set to obtain thirty- two responses. The analysis was evaluated by the effect of the variables on the responses obtained in the models and the relation between the size and the different values that were applied to the models.

The *treatments* were organized by considering the main factor of protein denaturation and this terminology was used to explain the thirty-two possible variables used on this study to determine the value of the responses (spray diameter, length and volume) in each particular model.

Economic analysis

The economic analysis was carried out in this study to determine the cost of building and operating a spray dryer over a specific range of operational conditions. Then, a comparative analysis was made to find out which model was the most economically efficient.

The cost estimation for the spray dryer was calculated following the economic considerations of Mujumdar [7]. This determined the total cost of a dryer taking into consideration the operation costs and the equipment costs, based on the quantity on the material and the dimensions of the diameter and the length.

The dryer unit was determined by the cost of wet product ($ / kg wet base) with Equation 5.

$$
\begin{aligned}
Cp &= \; CTtop\,Fs\,(1+Xs) \\
Ct &= eCeq + Cop \\
e &= i\,(1+i)N\,/\,(1+i)N\, -1 \\
Ceq &= D\,An'D + \alpha nZZn'z \qquad (5) \\
Cop &= hP\,Cetop + Z\,CzTop \\
hp &= qN'D\,(M+W') \\
W &= M\;\;(\,2D^2/\,4)\; + \;\pi DL)\,dx
\end{aligned}
$$

Size analysis and CFD simulation

A CFD simulation was carried out to investigate the fundamental flow behaviour in the spray dryers and their modeling. The simulation is performed using a mixed Eulerian/Lagrangian approach, in which the single single phase Reynolds Averaged Navier Stokes (RANS) equations are solved to determine the flowfield. This work used the responses given by the models to employ them as boundary conditions to simulate the spray dimensions and the CFD. The results of this simulations are shown in Figure 2.

DISCUSSION AND RESULTS

Depending on the modeling and boundary conditions of each author, variations in the dimensional parameters (diameter, length and volume) were obtained in the design models. When the models of Mujumdar [7], Suryanarayana [8], Gluckert [9] and Masters [10] were analyzed using the residence time and air velocity levels (high and low) as variables, only four responses of the possible thirty-two were obtained for the models of Mujumdar, Suryanarayana, and Gluckert [7-9] and one of thirty-two for Masters [10].

Model responses

The responses of the models were calculated by the combination of the five variables selected for this study in the two chosen levels (Figure 1). This procedure showed thirty-two responses for each dimensional variable (diameter, length and volume) for every model. The average responses of each model were compared to describe the behavior with the same operational conditions.

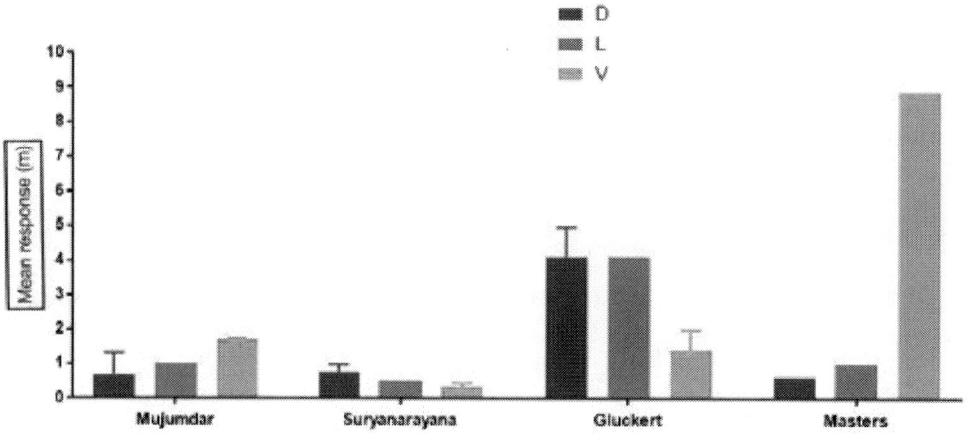

Figure 1. Model responses.
* "T" on top of the bars means average value conquered

The models of Mujumdar, Suryanarayana, Masters and Gluckert did not consider the denaturation conditions of whey protein for the designing of a spray dryer.

Models-comparison

The models comparison was developed with the analysis of the four models on two categories: sizing and modeling and cost estimation, this conditions were adjusted to analyze the scalability of the spray dryer of each model in terms of size of the chamber and cost estimation.

The sizing and modeling of the spray dryers were carried out in AutoCAD Inventor 2019, in this program was simulated the structure of the spray dryers according to the responses obtained on the variables analysis, the results demonstrated that exist appreciable differences on the sizes of the sprays dryers by each model.

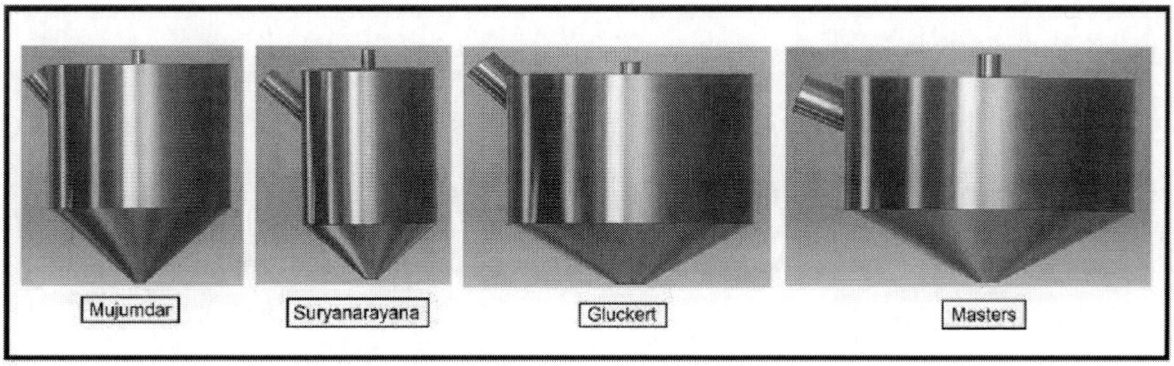

Figure 2. Size analysis

The cost estimation responses were calculated over a set of initial conditions for the 4 models expressed on the Table 2.

The model prices of the spray dryers varied from $6,417 USD to $14,803,215 USD. The lowest-priced option was the Suryanarayana [8] model with dimensional responses (diameter, length and volume) of 0.454 m, 0.313 m and 0.193 m3 respectively, and the highest-priced model was Gluckert [9] with dimensional responses (diameter, length and volume) of 5.301 m, 7.952 m and 0.912 m3 respectively.

Table 2. Cost estimation

Model	Responses (U.S. Dollars)
Mujumdar	$18,264 - $205,755
Suryanarayana	$6,417 - $91,540
Gluckert	$2,890,141 - $14,803,215
Masters	$27,711

CFD results

The results showed on the Figure 3 represent the flow patterns inside the spray chamber, the conditions proposed to carry out the simulation were the dimensions obtained on the responses of this work and the values of the temperature and the whey flow were adjusted according the conditions of non-denaturation presented on Table 1.

In this simulation (Figure 3) was possible to observe the changes that suffered the whey liquid particles inside the spray dryer, the particles traveled since the atomizer until the outlet hole, also is appreciable to see the the mass and energy transfer, as well as the the temperature profile in the middle of the spray dryer, represented by the mixture of colors in the center of the Figure 3.

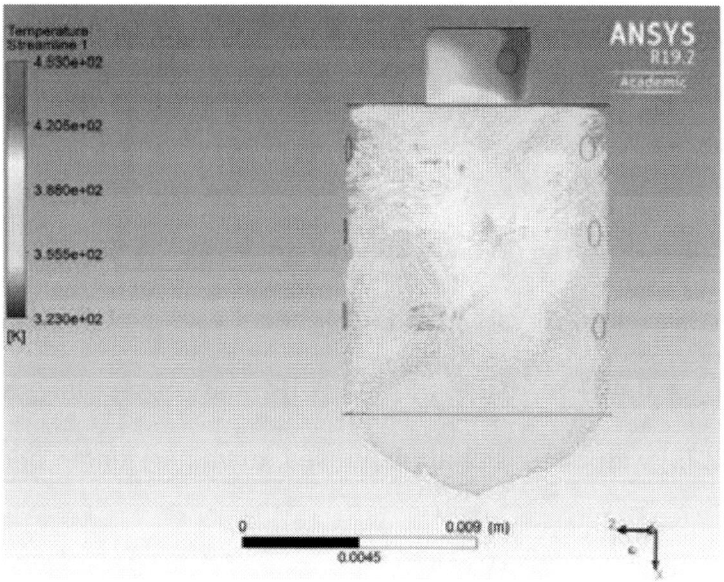

Figure 3. ANSYS CFD simulation

CONCLUSIONS

The models analyzed in this study presented different behaviors when evaluated over the same operational conditions. This is because the creator of each model used different mathematical correlations to describe the same phenomenon. This resulted in different responses in the spray dryer dimensional variables, (diameter, length and volume), that were analyzed by the simulations on ANSYS CFD who showed the behavior of the drop responses inside the spray chamber. Also in this work was studied the economic conditions for design a spray dryer to offer a guidance that allowed to analyze the cost estimation to develop the equipment. The analysis was carried with the key variables and responses adjusted in this research, obtaining a guidance to select a specific model design according the budget.

Acknowledgments

The authors gratefully acknowledge the Universidad del Atlántico in Barranquilla, Colombia for all the support and knowledge delivered. Thank you to Savannah Cayson for editing the writing of this article.

REFERENCES

[1] C. V. Morr, E. Y. W. Ha, Whey protein concentrates and isolates: Processing and functional properties. *Critical Reviews in Food Science and Nutrition*. **29** (2009) 431-476

[2] P. Schuck, R. Jeantet, B. Bhandari, X. D. Chen, Í. T Perrone, A. Fernandes de Carvalho, Recent advances in spray drying relevant to the dairy industry: A comprehensive critical review. *Drying Technol.* **34** (2016) 1773-1790

[3] M. Mezhericher, A. Levy, I. Borde, Spray drying modelling based on advanced droplet drying kinetics. *Chemical Engineering and Processing: Process Intensification*. **49** (2010) 1205-1213

[4] M. Mezhericher, A. Levy, I. Borde, Multi-Scale Multiphase Modeling of Transport Phenomena in Spray-Drying Processes. *Drying Technol.* **33** (2015) 2-23

[5] V. S. Birchal, L. Huang, A. S. Mujumdar, M. S. Passos, Spray Dryers: Modeling and Simulation. *Drying Technol.* **3** (2006) 275-279

[6] H. M. Lisboa, M. E. Duarte, M. E. Cavalcanti-Mata, Modeling of food drying processes in industrial spray dryers. *Food and Bioprod process.* **107** (2018) 49-60

[7] A.S. Mujumdar, *Handbook of industrial Drying.* (2014), 182-200

[8] A. Suryanarayana, *Mass Transfer Operations.* (2002), 571- 578

[9] F.A Gluckert, A theorical correlation of spray-dryer performance. *American Institute of Chemical Engineers*. **22** (1962) 713-724

[10] K. Masters, *Spray Drying Handbook*. (1979), 115-120

[11] M. Tunick, A. Thomas-Gahring, D. Van Hekken, S. Iandola, M. Singh, P. Qi, D. Ukuku, S. Mukhopadhyay, C. Onwulata, P. Tomasula, Physical and chemical changes in whey protein concentrate stored at elevated temperature and humidity. *Journal of Dairy Science*. **99** (2016) 2372-2383

[12] M.A. Haque, P. Aldred, J. Chen, C. Barrow, B. Adhikari, Comparative study of denaturation of whey protein isolate (WPI) in convective air drying and isothermal heat treatment processes. *Food chemistry*. **141** (2013) 702-711

536291 Molecular Thermodynamic Model for Aqueous Na^+–K^+–Mg^{2+}–Ca^{2+}–Cl^-/SO_4^{2-} Quinary Electrolyte Systems

Abstract:

Managing produced water is one of the major challenges in oil and gas production [1]. Produced water can be as much as seven times saline compared to sea water systems, and desalination of this water depends on novel material and process innovations. We present a comprehensive thermodynamic model aimed to support heat and mass balance calculations and predict mineral scale deposition of the saline water. The six major ions that control salinity of the water are: Na^+, K^+, Mg^{2+}, Ca^{2+}, Cl^-, and SO_4^{2-}. Current investigation focuses on establishing a thermodynamic model for the Na^+–K^+–Mg^{2+}–Ca^{2+}–Cl^- and the Na^+–K^+–Mg^{2+}–Ca^{2+}–SO_4^{2-} aqueous quinary electrolyte systems. To model the systems, we adopt electrolyte Non-Random Two Liquid theory (eNRTL) for electrolyte solutions [2] which describes concentration dependency of the solution nonideality with two binary interaction parameters for each water-electrolyte and electrolyte-electrolyte pair. Temperature dependency of each binary parameters is further correlated with a Gibbs-Helmholtz type expression with three temperature coefficients representing Gibbs energy, enthalpy, and heat capacity contributions. With the binary parameters and thermodynamic constants of the precipitating salts regressed from phase equilibria, calorimetric and salt solubility data, the model reliably calculates various thermodynamic properties i.e., salt solubility of the quaternary subsystems as well as quinary systems with temperatures from 273.15 to 473.15 K. The model is being extended to cover additional ions such as Sr^{2+}, Ba^{2+}, and HCO_3^- without losing inherent consistency and it should become a very useful tool for chemical engineering calculations.

Reference:

[1] A. V. García, K. Thomsen and E. H. Stenby, "Prediction of Mineral Scale Formation in Geothermal and Oilfield Operations Using the Extended UNIQUAC Model: Part I. Sulfate Scaling Minerals," Geothermics. 2005, 34(1): 61-97.

[2] Y. Song, C.-C. Chen, "Symmetric Electrolyte Nonrandom Two-Liquid Activity Coefficient Model," Ind. Eng. Chem. Res. 2009, 48:7788-7797.

530194 Numerical Investigation of Heat Transfer in Fixed-Bed Reactors Filled with Complex Particle Shapes Using CFD

Abstract:

Catalytic fixed-bed reactors with a low tube-to-particle diameter ratio are widely used in the chemical and process industry. They are characterized by a heterogenous packing morphology that lead to a strong interplay between local flow effects and the heat and mass transfer within the system. For that reason, simplified pseudo-homogenous models lead to an erroneous numerical description of that reactor type which is why more sophisticated numerical methods are needed.

Particle-resolved computational fluid dynamics (CFD) is a first-principle modeling approach that has been proven to be a reliable and predictive tool to analyze the flow, temperature and species field within fixed-beds in a spatially resolved manner as recently reviewed by *Jurtz et al.* This allows the method to be used as a design tool for new particle shapes to optimize the trade-off between a high active catalytic surface, a low pressure drop and a good heat transfer characteristic.

Within the scope of this presentation a fully automated workflow for the simulation of fluid flow and heat transfer in fixed-beds filled with cylindrical particles, Raschig rings and multi-hole cylinders will be presented. The different particle shapes will be compared regarding their thermal performance and advanced post-processing methods will be used to understand the complex interplay between particle orientation, convective heat transfer within the inner particle voids and the overall thermal performance. It will be shown how effective transport parameters can be extracted from the simulation results that afterwards can be used in simplified numerical models.

Literature
Jurtz, N., Kraume, M. and Wehinger, G. (2018). *Advances in fixed-bed reactor modeling using particle-resolved computational fluid dynamics (CFD).* Reviews in Chemical Engineering. Published: 2018/02/02. Available at: https://doi.org/10.1515/revce-2017-0059 [Epub ahead of print]

534211 Optimal Front-End Crude Scheduling for Refinery with Consideration of Proactive Unit Maintenance

Abstract:

The front-end crude scheduling (FECS) has been well studied and convincingly proven very effective for the scheduling of crude supply process [1,2]. It can be used to tackle crude movement problems from crude importing by vessels, unloading through berths, temporarily storing at port-side storage tanks, transferring via long-distance pipeline, blending at refinery-side charging tanks, to continuously feeding to refineries. Through FECS, optimal crude transfer schedules in terms of profitability and operability can be determined. In previous studies, most works were developed based on the assumption that all processing units were under normal working status without consideration of any possible abnormal performance. In reality, however, the performance of every processing unit will be inevitably decaying along with the processing time and load. Thus, regular maintenance operations have to be scheduled in advance to restore unit performing efficiency and more importantly to prevent potential unit failures [3,4]. Meanwhile, the unit maintenance schedule needs to be proactively integrated with the crude movement schedule to ensure the real operability as well as the minimization of the overall cost. Some research works did deal with maintenance issues by identifying post remedial strategies for unit failures [5,6]. However, the reactive response will unavoidably and stiffly disrupt the original schedule, resulting in higher operating expense and even infeasible operations. Therefore, proactive scheduling of regular unit maintenance is a promising opportunity to further improve and significantly benefit the crude supply process.

In this study, a novel continuous-time FECS model combined with proactive scheduling of unit maintenance has been developed. It involves a large-scale mixed-integer nonlinear programming problem, covering all crude movements from shipping all the way to the refinery processing. Through this model, the optimal schedule for both crude supply and unit maintenance operations will be simultaneously identified for the minimal operating cost. It can not only determine when and how to isolate the designated unit(s) from the process system for the maintenance preparation (i.e., how to discharge the leftover inventory in units), but also identify the best opportunity for the maintenance-completed unit(s) to rejoin the system for continuous production during the scheduling time horizon. In addition, real process requirements, including tank inventory, transfer flowrate, blending specifications are all taken into account to guarantee the obtained schedule is practical. The efficacy and performance of the proposed model has been tested and demonstrated through various unit(s) and maintenance scenarios.

References

[1] Zhang, S., & Xu, Q., (2015) Refinery continuous-time crude scheduling with consideration of long-distance pipeline transportation. Computer and Chemical Engineering, 75, 74-94.

[2] Li, J., Misener, R., & Floudas, C., (2012) Continuous-time modeling and global optimization approach for scheduling of crude oil operations. AIChE Journal, 58, 205-226.

[3] Amaran, S., Sahinidis, N., Sharda, B., Morrison, M., Bury, S., Miller, S., & Wassick, J., (2015) Long-term turnaround planning for integrated chemical sites. Computer and Chemical Engineering, 72, 145-158.

[4] Vieira, M., Varela, T., Povoa, A., (2017) Production and maintenance planning optimization in biopharmaceutical processed under performance decay using a continuous-time formulation: A multi-objective approach. Computer and Chemical Engineering, 107, 111-139.

[5] Zhang, S., Xu, Q., (2014) Reactive scheduling of short-term crude oil operations under uncertainties. Industrial & Engineering Chemistry Research, 53, 12502-12518.

[6] Zhang, S., Wang, S., & Xu, Q., (2015) A new reactive scheduling approach for short-term crude oil operations under tank malfunction. Industrial & Engineering Chemistry Research, 54, 12438-12454.

534540 Thermodynamic Modeling of CO_2 Absorption in Aqueous Amino Acid Salt Solutions with Symmetric Electrolyte NRTL Model

Abstract:

Development and large-scale deployment of renewable energy is the need of the hour, but it is still desirable to mitigate the increased CO_2 emissions as we make a transition from fossil fuel-based energy sources to renewable energy sources. In this perspective, many separation technologies have been designed for the large-scale removal of CO_2 from industrial sources. The most developed technology for CO_2 removal is the solvent based absorption using monoethanolamine (MEA)[1]. However, alkanolamines suffer from volatility, oxidative degradation and toxicity making them less suitable for solvent-based absorption[2,4]. Recently, aqueous solutions of amino acid salts (AAS) have gained considerable attention as alternative solvents for carbon capture because the inherent ionic nature of the system makes it non-volatile[2] which is advantageous at the stripper conditions. Also, their higher oxidative stability is favorable for oxygen-rich flue gas streams[4]. They also have comparable absorption kinetics to alkanolamines[2]. Thus, it is desirable to develop an extensive thermodynamic model of AAS-H_2O-CO_2 system to support process development and optimization. However, scarcity of extensive experimental data on thermodynamic aspects of the binary (AAS-H_2O) and ternary (AAS-H_2O-CO_2) systems[2] as well as the dissociation of salt to zwitterion, anion and cation[3] add to the complexity of modeling the liquid phase chemical equilibrium. Limited work has been reported for the experimental determination of CO_2 solubility in aqueous solutions of potassium salts of Glycine[4] (KGly), Sarcosine[2] (KSar), Proline[5] (KPro) and Taurine[6] (KTau). Prior models using extended UNIQUAC framework[2] and Deshmukh-Mather mehtodology[4] have been developed for potassium salts of sarcosine and glycine respectively.

In this study, we develop a symmetric electrolyte NRTL model to represent liquid phase non-idealities in conjunction with Redlich-Kwong equation of state to model vapor phase fugacity. Solubility data and pH data is used to identify eNRTL binary interaction parameters for one binary system (KSar-H_2O) and three ternary (KSar-H_2O-CO_2), (KGly-H_2O-CO_2), (KPro-H_2O-CO_2) systems. The models accurately represent the VLE behavior and liquid phase speciation in the three AAS-H_2O-CO_2 systems, covering a temperature range of 313-353 K and CO_2 loading of 0.2-1. The models can be further advanced upon the availability of more extensive experimental data on the absorption enthalpy for the three AAS-H_2O-CO_2 systems.

References:

1. Zhang, H.Que, C.-C.-Chen, Fluid Phase Equilibria 311 (2011) 67-75
2. E.Aronu, E.T.Hessen, T H.-Warberg, K.A.Hoff, H.F.Svendsen, Chemical Engineering Science 66 (2011) 2191-2198
3. -C.-Chen, Y.Zhu, Biotechnology Progress, Vol. 5 No. 3 (1989) 111-118
4. F.Portugal, J.M.Sousa, F.D.Magalhaes, A.Mendes, Chemical Engineering Science 64 (2009) 1993-2002
5. E.Majchrowicz, D.W.F.Brilman, Chemical Engineering Science 72 (2012) 35-44
6. S.Kumar, J.A.Hogendoorn, P.H.M.Feron and G.F.Veersteg, J.Chem.Eng.Data 46 (2001) 1357-1361.

525846 Plantwide Process Design with Automatic Column Optimization, Sequencing and Stacking Using a Rigorous Process Simulator

Abstract:

Column sequencing refers to the problem of determining the optimal column sequence of distillation columns for multicomponent separation. For example, for the separation of three components, ABC, there are two possible sequences: the direct split (A/BC followed by B/C) and the indirect split (AB/C followed by B/C). The number of possible sequences grows rapidly as the number of components to be separated grows. The column sequencing problem has been widely studied in the literature [1–11], although most researchers use shortcut equations (e.g. the Fenske-Underwood-Gilliland column design method) for the column design in order to reduce the computational load.

Column stacking refers to the process of adjusting the operating pressure of one or more columns so that the reboiler of one or more columns can be powered by condenser of another column. Depending on the temperature difference between the top and bottom of the columns and the temperatures of the utilities, three or more columns can be stacked on top of each other. This type of heat integration can significantly reduce energy consumption. Column stacking has also been widely studied in the literature [12–16], although again mostly using shortcut models.

Finally, it is well known that for many problems of practical interest, there is a tradeoff between the design of the reactor or reactor network and the cost of separation [17]. Specifically, for many problems of practical interest, selectivity decreases as per-pass conversion increases, so there is a tradeoff between selectivity losses at high conversion and separation cost (for recycle of unconsumed reactant) at low per-pass conversion. Thus in order to determine the optimal plant-wide process design, it is necessary to consider the design of the reactor or reactor network and the separation system simultaneously. For greater accuracy, it would be desirable to use a rigorous process simulator rather than shortcut column design equations.

In the present contribution, these problems are approached using a rigorous process simulator (Aspen Plus) coupled with a high-level procedural programming language (Matlab). The two programs are connected using a client-server architecture, in which Matlab is the client and Aspen Plus is the server. A process flowsheet consisting of a single distillation column is constructed in Aspen Plus, and this flowsheet is used to model each column in every sequence by changing the feed conditions. The reflux rate and boilup rate are manipulated (varied) in order to achieve desired specifications on the top and bottom product distributions. The optimal number of trays and feed tray location are determined by optimization using simulated annealing programmed in Matlab.

The code determines the optimal design of each column in every sequence and adds up the cost to determine the optimal column sequence. Once the optimal sequence is identified, the code automatically implements column stacking by adjusting the operating pressures of the columns within the limits of feasibility (given temperatures for the utilities and a minimum temperature driving force). Finally, this entire process (automatic column design, sequencing and stacking) is coupled with a model of a chemical reactor (in a separate Aspen flowsheet) to determine the optimal plantwide process design by adjusting reactor network design parameters (temperature and residence time) to find the optimal tradeoff between selectivity losses and separation costs using the optimized separation system design in each case.

References

[1] Nishida N.; G. Stephanopoulous; A. W. Westerberg. A review of process synthesis. AIChE Journal 1981, 27, 321-351.

[2] Tedder D. W; Rudd D. F. Parametric Studies in Industrial Distillation Part I. Design Comparisons. AIChE Journal 1978, 24, 303-315.

[3] Stephanopoulos G; Westerberg A. W. Studies in Process Synthesis–II. Evolutionary Synthesis of Optimal Process Flowsheets. Chemical Engineering Science 1976, 31, 195-204.

[4] Malone M. F.; K. Glinos; F. E. Marquez; J. M. Douglas. Simple analytical criterion for the sequencing of distillation columns. AIChE Journal 1985, 31, 686-689.

[5] P. B. Gadkari; R. Govind. Analytical Screening Criterion for Sequencing of Distillation Columns. Computer & Chemical Engineering 1988, 12, 1199-1213.

[6] M. Bezzina; L. Pibouleau; S. Domenech. A Statistical Approach to the Synthesis of Separation Sequences. Computer & Chemical Engineering 1989, 13, 651-660.

[7] Floquet, P.; L. Pibouleau.; S. Domenech. Separation Sequence Synthesis: How to Use Simulated Annealing Procedure? Computer. Chem. Eng 1994, 18, 1141-1148.

[8] An, W-Z.; Yuan X-G. A Simulated Annealing-Based Approach to the Optimal Synthesis of Heat-Integrated Distillation Sequences. Computers and Chemical Engineering 2009, 33, 199-212.

[9] F. Wang; Y. Luo; X. Yuan; A Formulation Methodology for Multicomponent Distillation Sequences Based on Stochastic Optimization. Chinese Journal of Chemical Engineering 2016, 24, 1229-1235.

[10] Boozarjomehry R. B.; A. Pourahmadi Laleh; W. Y. Svrcek. Automatic Design of Conventional Distillation Column Sequence by Genetic Algorithm. The Canadian Journal of Chemical Engineering 2009, 87, 477-492.

[11] Boozarjomehry R. B.; A. Pourahmadi Laleh; W. Y. Svrcek. Evolutionary Design of Optimum Distillation Column Sequence. The Canadian Journal of Chemical Engineering 2012, 90, 956-972.

[12] Andrecovich, M. J.; A. W. Westerberg. A Simple Synthesis Method Based on Utility Bounding for Heat-Integrated Distillation Sequences. AIChE Journal. 1985(b); 31: 363-375.

[13] Andrecovich, M. J.; A. W. Westerberg. An MILP Formulation for Heat-Integrated Distillation Sequence Synthesis. AIChE Journal 1985(a), 31, 1461-1474.

[14] Andrecovich, M. J.; A. W. Westerberg. Utility Bounds for Nonconstant QΔT for Heat-Integrated Distillation Sequence Synthesis. AIChE Journal 1985(c), 31, 1475-1479.

[15] C. A. Floudas; G. E. Paules IV. A Mixed-Integer Nonlinear Programming Formulation for the Synthesis of Heat-Integrated Distillation Sequences. Computers & Chemical Engineering 1988, 12, 531-546.

[16] Jaroslav, J.; Radim Ptacnlk. Synthesis of heat integrated rectification systems. Comput. Chem. Engng 1988, 12, 427-432.

[17] Douglas, J. M. Conceptual design of chemical processes. New York, McGraw-Hill, 1988.

538224 Optimal Design of Dynamic Experiments for Pilot Plants for CO_2 Capture

Abstract:

The systematic design of dynamic experiments (DoDE), especially for pilot plants for CO_2 capture, is an area that has received little attention in the literature. Since pilot plant test runs require significant resources, Dynamic experiments can easily contain over 1500 data points with meaningful information gathered in a day, while conventional steady-state design of experiments yield 2 or 3 data points. Thus, an optimal DoDE is developed and implemented in a pilot plant for CO_2 capture.

The objective of the DoDE is to maximize the observability of model parameters within a given time constraint for the test runs. For DoDE, it is required that the inputs signals are persistently exicted so that the observability of the model parameters can be ensured. For real-life implementation of the inputs signals, the designed signal should be "plant-friendly". Plant-friendliness of the input signals ensures that they do not lead to unacceptable changes in product quality and controller set-points that could cause "wear and tear" on the process equipment. Furthermore, the input signals should not lead to unsafe operation of the plant. For designing plant-friendly input signals, one needs to also consider the crest factor, which is the peak amplitude in a signal waveform, in addition to the persistence of excitation. If these properties are not considered, it can lead to signals that are practically unacceptable due to the signal variability, frequency content (harsh changes), amplitude (designed values cannot be achieved at implementation) and waveform (some signal forms may not be implementable in a given control system).

The dynamic model of the pilot plant used for DoDE was developed as part of the U.S. Department of Energy's *Carbon Capture Simulation Initiative* (CCSI)[1]. The model includes rigorous thermodynamic[2] and transport models[3] for the MEA-H_2O-CO_2 system, along with a rigorous mass transfer model that was developed by simultaneously estimating the parameters for the interfacial area, reaction kinetics, mass transfer coefficients and diffusivity models by concurrently using the data from wetted wall column and packed column experiments[4]. A pseudo-random binary sequence (PRBS) is designed as the input signal for DoDE due to its efficiency in obtaining sufficient spectral content[5]. However, due to the long sequence size, the PRBS signal can be time-consuming and prohibitive for large-order systems. Therefore, a Schroeder-phased input signal, which is a multisine signal, is also designed. This signal has similar characteristics to the PRBS yet can be implemented within reasonable time for higher-order systems. The design also ensured plant friendliness and could be successfully implemented in the Pilot Solvent Test Unit (PSTU) at the National Carbon Capture Center in Wilsonville, Alabama. The transient data are used to solve a dynamic data reconciliation and parameter estimation problem. It is observed that the parameters estimated from dynamic data collected over only a day are nearly equivalent to those obtained from the steady-state data collected over several weeks.

References

1. Miller, D. C., Syamlal, M., Mebane, D. S., Storlie, C., Bhattacharyya, D., Sahinidis, N. V., Sun, X. Carbon capture simulation initiative: a case study in multiscale modeling and new challenges. *Annual review of chemical and biomolecular engineering* **2014**, *5*, 301-323,.
2. Morgan, J. C., Chinen, A. S., Omell, B., Bhattacharyya, D., Tong, C., Miller, D.C. Thermodynamic Modeling and Uncertainty Quantification of CO_2-Loaded Aqueous MEA Solutions. *Chemical Engineering Science* **2017.** *168*, 309–324
3. Morgan, J. C., Bhattacharyya, D., Tong, C., Miller, D.C. Uncertainty Quantification of Property Models: Methodology and Its Application to CO_2-Loaded Aqueous MEA Solutions. *AIChE Journal* **2015**. *61*, 1822-1839.
4. Chinen, A. S., Morgan, J. C., Omell, B., Bhattacharyya, D., Tong, C., Miller, D.C. Development of a Gold-Standard Model for Solvent-Based CO2 Capture. Part 1: Hydraulic and Mass Transfer Models and Their Uncertainty Quantification. *Industrial & Engineering Chemistry Research* **2018**, (submitted)
5. Hjalmarsson, H. (2014). *Experiment Design and Identification for Control.* Encyclopedia of Systems and Control.

530117 Review of the Different Type of Flow Elements and Technologies Suitable for Demo Scale Plants, Pilot Plants, and R&D Equipment with Installed Cost Comparison

Abstract:

The objective of this presentation is to review the various type of flow elements and the feasibility for small scale modular plants, pilot plants and laboratory units. As a manufacturer of modular pilot plants, Zeton has been involved in a wide variety of pilot plant design and construction, and has first-hand experience with the advantages and disadvantages of selecting different type of flow elements for pilot plants, demo-scale plants and research units. Often the uniqueness of small line sizes in tubing and piping are not carefully considered when deciding on the correct technology and suitability. Typical technologies that can be used in well-known, well characterized processes are not always suitable with new processes. Factors such as cost, space minimization, and pilot plant life are often not considered at the same time. This paper will review several flow technologies with examples comparing cost and space differences where practical small scale modular plant and pilot plant design practices have been utilized in meeting process objectives.

533939 Innovating Chemical R&D Processes By Treating Them As Metaphorical "Chemical Processes"

Abstract:

The goal of this talk is to frame the Chemical R&D process itself as a type of "chemical process". This framing enables dramatic improvements in throughput, value of products and services for customers, speed from idea to market, improved sustainability, and more. In this talk we analyze the innovation process according to "chemistries" (e.g., idea catalysts and generation, idea transfer concepts), streams (e.g., desired outputs, inputs, recycles, purges), and unit operations (e.g., idea generation "reactors" and idea evaluation "separators"). This perspective enables the analysis and measurement of R&D, including analysis of dynamics, providing a coherent philosophy for innovation processes that can significantly improve the flow of valuable R&D innovations.

527050 Hydrodynamics Characteristics of a Stirred Tank Provided with Angle-Mounted Impellers Using Computational and Experimental Approaches

Abstract:

Most industrial stirred tanks and reactors are provided with a centrally mounted impeller to ensure adequate mixing and with wall baffles to prevent the swirling effect that would otherwise occur in an unbaffled vessel. However, an alternative approach to reduce swirling and promote mixing in an unbaffled tank is to place the impeller asymmetrically and mounted on an angled shaft. This type of configuration is common in a number of industrial operations and especially in the pharmaceutical industry, since the absence of baffles reduces the potential for contamination and ensures thorough cleaning of the vessel between batches while introducing adequate top-to-bottom liquid recirculation of the vessel content and effectively eliminating swirling effects. Although industrially relevant and relatively common in the industrial practice, only limited information is available on the operation of tanks provided with angle-mounted impellers and even less is known of their hydrodynamics. Therefore, the objective of this work was to study the hydrodynamics of a scaled-down version (diameter: 316 mm) of a tall industrial vessel with an elliptical bottom and provided with two angle-mounted (by 5Â° off the vertical) A310 Lightnin impellers under different operating conditions in order to determine the mixing characteristics of the industrial system. Both Particle Image Velocimetry (PIV) and Computational Fluid Dynamics (CFD) were used to quantify the hydrodynamics and provide guidance on the optimal operation of these industrial system. The PIV apparatus was used to experimentally determine the velocity flow field in water at different agitation speeds so as to validate the CFD predictions. A multiple reference frame (MRF) computational approach and turbulence models based on the k-Îµ method were applied for all cases in the CFD simulations. The results revealed substantial agreement between the CFD predictions and the PIV experimental results. A complex flow in the vessel was observed. Due to the axial pumping action of the impellers, a downward flow impinging the vessel bottom was created resulting in the splitting of the flow. A portion of the zone near the bottom of the vessel, which is the preferential location for the sedimentation of settling solids in the liquid, was dominated a less well-mixed zone formation. Additionally, increasing agitation speeds had no significant effect on the presence of that zone.

530926 Experimental and CFD Study of Mixing Two Fluids of Different Properties and Its Application in Biological Drug Product Manufacture

Abstract:

During biological drug product manufacture (solution in syringe or vial), concentrated drug substance in solution, typically with high viscosity, is either used as is or diluted with buffer to desired concentration. The high viscosity drug substance is amenable for subcutaneous administration but is difficult to administrator through intravenous injection (IV) without dilution. While subcutaneous administration can be conducted at home by a non-healthcare professional, intravenous injection must be administrated by a healthcare professional. Therefore, highly concentrated drug substance with high viscosity is desired at the starting point of drug product process to allow flexibility between subcutaneous administration (used as is) and intravenous injection (used after dilution). Efficient dilution of highly concentrated drug substance can be achieved at an appropriate mixing speed in a given mixer. However, the max mixing speed is set by the sensitivity of the biological molecules to shear and interfacial stress. A speed too fast may raise stability concerns. Hence, an optimal mixing condition balances between efficiency and product stability.

In this study, sucrose solution of various concentrations is mixed with water or water like liquid to mimic the two-fluid mixing system in biological drug product formulation process(Sucrose solution: drug substance, water: buffer). The impacts of initial concentration of sucrose and impeller speed on the mixing time are investigated using both experiments and computational fluid dynamic (CFD) simulation at three different scale (impeller bottom driven) mixing tanks, 250ml, 10L and 400L respectively. The experimental results in 250ml and 10L tank show that there exists a minimum mixing speed at which two fluids homogenous mixing occurs. It is also found that the impact of the initial concentration of sucrose solution on mixing time decreases at the mixing speed increases. To scale up two fluids mixing system using the traditional dimensionless number like Reynold number or tip speed is challenging due to the complexity of two-fluid mixing. A CFD two-fluid mixing model recognizing the difference of viscosity and density between the two fluids is developed for scale up. The model captures the characteristics of two-fluid mixing as described in experiments. After validated by the experimental at 250ml and 10L tanks, the model is applied at a 400L mixing tank. The model suggests key mixing operation parameters based on the liquid volume in the tank. The suggested parameters work as expected from the model.

* Presenter.

533597 Mini Continuous Stirred Tank Reactors (mini-CSTR) for Cell and Tissue Culture Applications

Abstract:

An ideal cell culture system will provide a well-controlled, homogeneous, and steady environment for cells and tissues. For instance, well-controlled steady states would greatly benefit organ-on-chip experiments, stem cell culture, and tissue propagation (among other relevant biomedical applications). However, at present, no continuously stirred mini-reactors are commercially available for lab-scale culture applications.

We are developing simple, low-cost, and user-friendly miniaturized continuously stirred tank reactors (CSTRs) for biomedical and biotechnological agitations. These well-mixed mini-CSTRs will enable cost-efficient continuous culture at small scales. We use high resolution 3D-printing to fabricate these CSTRs and an Arduino platform to measure and control key parameters, such as temperature and pressure, in small portable incubators. Nutrients are fed by syringe pumps, and well-controlled low-speed (benign) agitation is provided by a custom-made magnetic system. Since the reactor behaves as a well-mixed reservoir, all bulk-liquid concentrations can be measured at the outlet stream, thereby greatly reducing the need for intrusive instrumentation. Here we present an exhaustive validation of the mixing performance of this system, including results from computational fluid dynamic experiments (CFD) at different agitation speeds. In addition, we present preliminary results of the use of this culture platform in two model applications: (a) the extended culture of breast cancer spheroids, and (b) the culture of Chinese Hamster Ovary Cells (the warhorse for biopharmaceutical production) for continuous production of biopharmaceutical compounds.

533915 Power Dissipation and Power Numbers for a Retreat-Blade Impeller in Pharmaceutical Mixing Tanks and Reactors Using an Experimentally Validated Computational Approach

Abstract:

Mixing in stirred vessels is a very common operation in many pharmaceutical processes. Glass-lined tanks and reactors are especially commonly used in the in pharmaceutical industry for the synthesis of Active Pharmaceutical Ingredients (APIs) because of their corrosion resistance, cleanability, and reduced contamination levels. In such cases, a retreat-blade impeller with a low impeller clearance off the tank bottom is commonly used in and the reactors are cylindrical and provided with a torispherical vessel bottom and equipped with a single baffle and a retreat-blade impeller (RBI). The power, *P*, dissipated by the impeller is critical to any mixing process since mechanical energy dissipation is needed to homogenize the vessel content, disperse immiscible phases, suspend solids, increase mass transfer, and, in general, produce the desired mixing level. *P* depends not only on the type of impeller used, agitation speed, and physical properties of fluid but also on the geometry of the system, including the location of the impeller in the mixing vessel. Despite their common use, little information is available on the hydrodynamics of RBIâ€™s is such systems and their mixing performance, including their power dissipation. In this work the power, *P*, dissipated by a RBI, and the non-dimensional Power Number, Po, in different fluids were obtained both computationally via Computational Fluid Dynamics (CFD) and experimentally, using a strain-gage based torque measuring device, for different Reynoldâ€™s Number values in a 61-Liter, scaled-down version of a typical glass-lined reactor for the synthesis of pharmaceutical APIs. Additionally, P and Po were obtained for a fully baffled system (four rectangular baffles) and for an unbaffled system. *P* and Po were obtained for a very large range of the Reynolds Number, Re (1<Re<1,000,000). Po was found to vary significantly with baffling type and Re. Excellent agreement was obtained between the CFD-predicted and the experimentally obtained values for Po. Correlation equations relating Po with Re using baffling type as the parameter were additionally obtained. These equations can be used by industrial practitioners to obtain Po and hence P for their system and thus help quantify and optimize pharmaceutical mixing processes typically involving the synthesis of APIs in this type of reactors.

540229 Accelerating Development and Intensification of Chemical Processes

Abstract:

Automated optimization and screening of production candidates along with process intensification offer opportunities for faster development and more efficient production for chemical industries. It is especially the case for those industries with many diverse products, such as specialty chemicals and pharmaceuticals. This presentation starts with opportunities in the early discovery/development phases enabled by automated screening and optimization of chemical reactions in microliter-scale droplets. The technology facilitates optimization across continuous variables (e.g., temperature, concentration, residence time) as well as discrete variables (e.g., catalyst species, base, and solvents) without requiring system depressurization or reconfiguration. Moreover, kinetic models can be extracted for subsequent scaling to process conditions.

As the next step in the development process, a robotically configurable chemical synthesis platform demonstrates automated, optimized multiple step continuous synthesis of molecular targets according to a predetermined recipe. Multistep synthesis occurs at elevated temperatures and pressures to enhance reaction rates, and the resulting residence times are on the order of minutes, in contrast to the multiple hour-long syntheses typically needed for batch processing. Process intensification is exemplified by on-demand synthesis of common pharmaceuticals, e.g., ciprofloxacin, in a plug-and-play, manually reconfigurable, refrigerator-sized manufacturing platform of integrated unit operations. CNC machining and 3D printing are used to realize the miniaturized process equipment described throughout the presentation.

The above advances in laboratory automation and translation to small-scale manufacturing promise to accelerate development, but planning chemical synthesis often requires considerable time and effort that can be reduced through computer-aided organic synthesis. Machine learning opens new approaches to retrosynthetic planning including the underlying challenges in reaction validation and forward reaction prediction. An overall workflow integrates the developed techniques for retrosynthesis and forward prediction to predict an ordered list of reaction paths to a given molecular target from purchasable starting materials via feasible reaction steps. The prospect for integrating these technologies into reaction engineering are discussed and shown through case studies.

534111 Purification of Exosomes Using Tangential Flow Filtration

Abstract:

Background: Increasing interest is being paid to exosomes, small membrane bound particles produced by cells, particularly due to their ability to carry proteins, RNA, and DNA. Specifically, exosomes are thought to play a large role in cancer, specifically as biomarkers and therapeutic agents [1]. However, one major difficulty in exosome research is the isolation and concentration of exosomes. The current standard method of exosome isolation is ultracentrifugation. Significant difficulties exist in ultracentrifugation of exosomes, primarily recovery. Ultracentrifugation has been shown to produce very low isolation concentrations and altered size distributions when using liposomes as a model system[2]. Additionally, ultracentrifugation is a lengthy process, requires expensive equipment, and possibly damages exosomes due to the high speeds[3]. Specifically, the high g-forces in ultracentrifugation are believed to cause the fusion of exosomes. When compared to other commercially available separation kits, ultracentrifugation has been seen to result in larger particle sizes, which is thought to be a result of exosomes fusing with other exosomes, contaminants, and proteins[4]. A high degree of variability has also been observed between different rotors. When the same conditions are run on different rotors, very different concentrations and size distributions are observed in the final concentration solution[5]. Based on these issues, a better method of isolation of exosomes is necessary. Any method should exhibit sizes consistent with the initial distribution, and should not show a drastic loss of sample. We propose the use of tangential flow filtration (TFF) which has been previously used for protein sepearations[6].

Methods: Cell supernatant is isolated from various cell lines of interest that are known to produce exosomes, HEK293, MDA 231, and CHO. After the supernatant is collected, it undergoes two benchtop centrifuge steps that are a part of the conventional ultracentrifugation process. The first, 300g for 5 minutes removes the cells, and the second 2000g for 30 minutes removes most cell debris. The supernatant after these two centrifugation steps is then loaded for filtration, with a small sample removed to allow for comparison to initial levels. Filtration is performed using a polysulfone hollow fiber filter with varying pore sizes, 10 kD, 50 kD, and 500 kD (Spectrum Labs). The supernatant is pumped through the filter, at 24 mL/min using a peristaltic pump. After passing through the filter, the retentate is returned to the initial sample container for recirculation through the system. The filtration continues until the retentate volume reaches around 3 to 5 mL remaining. The retentate of filter contains the concentrated exosomes, and the permeate media and other small components and is discarded. When collecting the retentate, the filter is flushed with 1 mL of PBS to remove exosomes that might still be on the filter. After the filtration, the initial sample, the retentate, and samples from the permeate are then tested using Nanoparticle Tracking Analysis (Nanosight) to determine the concentration and size distribution of the exosomes. We then calculate the total number of particles in both the initial and final samples to see the recovery.

Results and Discussion: Our initial trials were all run using the 10 kD filter, further experimentation is needed on the 50 and 500 kD filters. Additionally, HEK293 cells were not used in these initial runs. For the MDA 231 cells, we started with 33 mL of supernatant, which is measured to have a total particle count on the order of 10^{11} particles, with some variability between individual harvests. For most runs, the final pellet is measured to be around 6 mL, and has a similar total particle count, on the order of 10^{11} total particles. In these initial runs we are seeing approximately a 5 fold concentration of the exosomes in the sample, while reducing the volume by approximately an order of 5. Additionally, the particle size distribution is far more similar to the initial distribution than is seen in ultracentrifugation. When running the permeate samples, the concentration of particles is far lower than in the retentate, indicating very few exosomes are going through the filter. There is still a degree of variability seen in the filtration results, though far less severe than with ultracentrifugation. While most samples show a final particle count very close to the initial sample, an occasional trial shows a concentration significantly lower than the initial, or slightly higher than the initial.

The CHO samples exhibit far more variability than the MDA 231 samples. These samples started with more volume, around 45 mL, and were concentrated to around 5 mL. The final particle count is greater than the initial count consistently across the CHO samples, in many cases significantly higher. Due to the higher initial volume, we have been able to concentrate the CHO exosomes to a final concentration of around 10^{12} particles/mL. Size distribution is far more consistent to the initial sample compared to ultracentrifugation.

Conclusion: Ultracentrifugation as a method of purifying exosomes has significant drawbacks, from the cost and length of operation, to low recovery and size inconsistencies. Tangential Flow filtration has the potential to be a better method for the concentration of exosomes from cell culture supernatant. Initial runs have shown that a far greater amount of the initial exosome concentration is recovered after TFF than after ultracentrifugation, and that the size distribution after TFF is far more indicative of the initial population than an ultracentrifuged sample. There are still concerns that need to be addressed with future work. Ultracentrifugation contains an additional step that has not yet been accounted for, a 10,000g centrifugation step that is intended to remove apoptotic bodies. Further runs will be performed using larger filters, where the exosomes should pass through as the permeate, and apoptotic bodies should remain in the retentate. This will likely impact the recovery of exosomes, as some might remain in the retentate. Other concerns include shear stress on the exosomes causing damage or deforming them enough to pass through the small membrane, and the length of time required to reach the desired volume. Back pressure applied plays a role in these concerns, a higher back pressure allowed us to reach the concentrated volume faster, but it could cause more exosomes to be squeezed through the pores of the membrane leading to lower recovery. Overall, initial results indicate that there is a potential of the TFF system being a promising way to concentrate exosomes, alleviating some of the issues associated with ultracentrifugation.

[1] Guo, W., Gau, Y., et al. 2017, "Exosomes: New players in cancer", Oncol Rep. Vol 38, pp. 665-675.

[2] Lane, R., Korbie, D, et. Al. 2015, "Analysis of exosome purification methods using a model liposome system and tunable-resistive pulse sensing", Scientific Reports, Vol 5, article number 7639. doi:10.1038/srep07639

[3] Li, P. Kaslan, M. et al. 2017 "Progress in Exosome Isolation Techniques", Theranostics. Vol 7, pp. 789-804.

[4] Helwa, I. Cai, J. et al. 2017 "A Comparative Study of Serum Exosome Isolation Using Differential Ultracentrifugation and Three Commercial Reagents". PLOS ONE doi:10.1371/journal.pone.0170628

[5] Livshits, M. Khomyaova, E. et al. 2015. "Isolation of exosomes by differential centrifugation: Theoretical analysis of a commonly used protocol". Scientific Reports, Volume 5, article number 17319 doi:10.1038/srep17319

[6] van Reis, R., Leonard, L. et al. 1991. "Industrial scale harvest of proteins from mammalian cell culture by tangential flow filtration". Biotechnology and Bioengineering, Vol 38, pp 413-422.

523035 Model-Based Optimization and Design of Continuous Chromatography for Protein Capture

Abstract:

The petrochemical, chemical and food industries have practiced continuous manufacturing for many years. However, biotechnology is still late to embrace continuous processing, especially for downstream processing. Recently, multi-column periodic counter-current (PCC) chromatography has been proposed as continuous capturing technology to improve the process productivity and resin capacity utilization, as well as reduce buffer consumption and equipment footprint, which is applying for monoclonal antibody capture with Protein A affinity resin. Due to the complexity of continues chromatography, some mathematical framework should be developed to aid the optimization and process design.

To better understand and describe continuous chromatography process, a fully mechanism model was built for fitting and predicting breakthrough curves as well as optimizing the operation conditions. An integrated approach was developed to combine the Equilibrium Dispersive Model of column, General Rate Model of resin, as well as the Equilibrium Adsorption Model for protein adsorption. Firstly, breakthrough curves at different flow rates and loading concentrations are fitted to obtain the model parameters, including pore diffusion coefficient, mass transfer coefficient and so on. Then, multi-column adsorption pattern and continuous separation are evaluated and predicted through these parameters under varying operation conditions. Finally, the working window would be determined with the design strategy, considering the process productivity and capacity utilization. In addition, the performance of continuous chromatography would be compared with traditional batch process, and the advantages and limitation of continuous processing would be discussed. The model-based approach developed were tested with different continuous multi-column modes, including twin-column CaptureSMB, 3/4-column PCC from GE Healthcare and multi-column BioSMB from Pall. The interactive interface was also developed to easily input the data and visually output the results. The results indicated that model-based approach could aid the process design and enhance both the development and application of continuous chromatography.

This work was supported by the International Science & Technology Cooperation Program of China and National Natural Science Foundation of China.

530329 Sterile Filtration of Oncolytic Viruses Using Novel Nanoporous Silicon Nitride Membranes

Abstract:

Oncolytic viruses (OVs) are a growing class of bio-therapeutics that have attracted significant interest for cancer immunotherapy and are expected to be worth over $1B by 2026. In 2016 alone, there were over 40 clinical trials (phase I, II, and III) conducted, both with OVs on their own and in combination with other cancer therapies. Current Good Manufacturing Practice (GMP) standards for the manufacturing of OVs rely on sterile filtration processes using microporous polymeric membranes to remove microbial contaminants. While this is necessary to ensure the purity and safety of the formulation, sterile filtration can also lead to reduced yield and increased process costs due to entrapment of viruses within the high internal surface area and broad pore distributions of the membrane; new process technologies are required to help optimize the production of OVs.

Nanoslit silicon nitride (NSN) membranes are an emerging technology of considerable potential to address this gap. Fabricated using advanced silicon processing techniques, these ultrathin (<400 nm) membranes have precisely controlled pore size (200-500 nm) and geometry (slit or circular). It is hypothesized that the unique properties of NSN membranes will allow it to significantly reduce the titer loss during sterile filtration while also being amenable to filtration at higher fluxes and lower pressures as compared to traditional membranes. We set out to demonstrate that the novel NSN membranes would enable sterile filtration of viruses while also having titer loss less than 10% (given starting load of ~10^{10} PFU/mL) and filtration capacity greater than or equal to commercial membranes (PVDF, PES). The NPN membranes were integrated into custom-built dead-end micro-scale filtration cells and studied under constant flux filtration, while a wide range of variables were monitored and characterized for their effect on performance including membrane properties, operational parameters, and feed composition. The filtration studies were performed using an attenuated strain of the oncolytic Maraba virus, a Rhabdovirus with a unique bullet shape (170 nm x 70 nm) which is currently undergoing clinical trails and shows promise as both an immunotherapy vaccine vector and oncolytic agent. Appropriate theoretical frameworks were used to analyze the experimental results in terms of the underlying bio-physical properties of the OVs and associated impurities (e.g. DNA, host cell proteins).

537865 Docking Simulations to Predict Binding Performance of Affinity Ligands for Purification of Butyrylcholinesterase

Abstract:

An ever-increasing demand for affordable biotherapeutics manufactured under stringent regulatory guidelines is driving the development of innovative, more efficient, lower-cost downstream purification technologies. The use of an affinity separation step in downstream purification can significantly reduce the number of process steps, buffer use, production time and overall costs. Affinity resins using antibody and other protein ligands can be prohibitively expensive. Synthetic peptide ligands offer a robust, lower-cost alternative to biological affinity ligands. This work illustrates the use of computational tools in identification, development and characterization of peptide based ligands for the purification of butyrylcholinesterase - an enzyme important for its ability to degrade cocaine, hydrolyze ester-containing drugs, and as an antidote to various nerve gas and organo-phosphorous compounds. This paper describes the use of High Ambiguity Driven Molecular Docking (HADDOCK) software to identify peptides that will bind to the protein through its active site. Docking calculations were performed for *in-silico* peptide libraries of three amino acids. Using characteristics of reported butyrylcholinesterase inhibitor molecule(s) as a guideline, peptides containing aromatic residues were prioritized for creating the peptide library. Residue cysteine forms cystine through a disulfide linkage, while methionine forms methionine sulfoxide (MetO) by addition of oxygen to its sulfur atom and were not considered while designing the library. Out of a total possible 8000 peptides, a biased library of 2256 peptides was designed with peptides containing one or more aromatic residues. Based on the available description of the active site of human butyrylcholinesterase, a targeted docking of these chosen 2256 peptides to the active site of butyrylcholinesterase was performed. A three-stage approach was adopted for this docking simulation â€“ rigid, semi-flexible and finally water refined fully flexible docking. The docked structures were grouped in clusters based on root-mean-square-distance (RMSD) and scored using Firedock software to calculate free energies of binding. Procainamide, a known ligand to human butyrylcholinesterase, was also docked using the same protocol and its binding energy was used as a reference. Binding characteristics of a few chosen peptide ligands were investigated by synthesizing them on chromatographic resins via Fmoc based solid phase peptide synthesis. Experimental binding results for synthesized peptides were compared to predictions made by HADDOCK.

538173 Computational Design of Peptide Ligands for the Bioseparation of "Fab" Antibody Fragments

Abstract:

An exciting alternative therapeutic to monoclonal antibodies are antigen-binding fragments of antibodies, "Fab", which retain the target specificity of their parental MAb but are easier and cheaper to produce, and have higher tissue penetration and quicker uptake. A key issue in Fab production is separating it from the cell cultures in which it is made during bio manufacturing. The industry standard to capture Fab from cell culture fluid is affinity chromatography using the affinity ligand Protein L which itself has a number of drawbacks including very high cost, low chemical stability, modest binding capacity, and potential immunogenicity. We describe a novel computational approach to discover new short peptides that can bind Fab specifically and selectively and then release it in eluting conditions in affinity chromatography columns thereby avoiding the need for harsh acidic conditions that denature antibodies. Two of the designed binders from the computational algorithm, were synthesized on Toyopearl Amino 650 M resin , tested for binding and purification of IgG from a cell culture harvest, and found to bind strongly and selectively .

540183 Downstream Processing in Biomanufacturing: Multimodal Chromatography, Affinity Precipitation and Integrated Bioprocessing

Abstract:

This talk will present recent developments in our lab in three areas of downstream bioprocessing; multimodal chromatography, affinity precipitation and integrated bioprocessing. Targeted experiments with different domains of mAbs with varying properties, biophysics, thermodynamics and molecular simulations will be used to shed light on the importance of protein surface cluster properties and multimodal ligand hydration and flexibility on creating selective separations. Hierarchical clustering is employed to examine preferred ligand conformations and the impact of water and dewetting phenomena in multimodal systems. NMR and molecular simulations will be used to examine the molecular basis for multimodal ligand binding to the FC region of mAbs both in solution and immobilized on SAM surfaces. In addition, pH and ligand chemistry will be employed to shift domain dominance in mAb multimodal chromatography from (Fab)2 interactions to Fc binding. Affinity precipitation using smart biopolymers for the simultaneous recovery and purification of both mAb and non-mAb biologics will then be presented. The use of ELP-Z for mAb purification as well as the development of efficient processes based on this approach will be presented with industrial feed stocks. The extension of this approach using ELP-affinity peptides as well as the application to a new class of biologics will also be discussed. Finally, results will be given on a novel approach for the rapid development of integrated downstream biomanufacturing processes for biological products from Pichia Pastoris. This approach employs linear gradient screens in concert with high-throughput uplc analytics and custom software to identify sequences of orthogonal chromatography steps likely to recover and purify the product while also being amenable to integrated processing and a continuous flowpath. To illustrate the power of this novel approach, the rapid development of 3-step purification processes for three commercial biological products will be shown.

537421 Application of Mechanistic Models for the Digital Design and Online Control of Pharmaceutical Processes

Abstract:

Mechanistic models are becoming more commonly applied for Research and Development in the pharmaceutical sector. Traditionally, the output from this activity, namely a validated mechanistic model, which is capable of quantitatively predicting the behaviour of the various Critical Quality Attributes (CQA) for typical batch or continuous pharmaceutical processes for a wide range of Critical Process Parameters (CPP). However, these tools are almost exclusively employed in an offline manner currently, primary aimed at assessing process robustness and variability, with very little subsequent online application of the model for control or soft sensing purposes.

Traditionally, online control companies utilize statistical models that require a significant level of tuning and verification, with the online full-scale plant, using Pseudo-Random Binary Sequences (PRBS) to vary the process parameters and observed the process responses. This requirement for the application of rigorous Model Predictive Control (MPC) using statistical model approaches commonly leads to reduced appetite for uptake of the technology in the Pharmaceutical sector. These online trials for tuning the MPC may result in off-specification productions, potentially leading to significant losses in productivity, as the controllers pushes the process to observe responses at extreme points. However, most of these drawbacks can be overcome by the integration of mechanistic models, developed using laboratory scale data with MPC system, such as that offered by Perceptive Engineering, namely PharmaMV.

In this work we outline, the application of an advanced process modelling tool, namely gPROMS FormulatedProducts, to describe a number of pharmaceutical processes. The process model and the mechanistic model kinetic parameters were validated using process data gathered from the literature and from lab-based experiments. The lab-based process was subsequently used to predict the behaviour of the full scale production scale. In order to make the model predictive, some refinement of the kinetic parameters for secondary nucleation was required using minimal experimental data from the typical plant runs.

The mechanistic model was integrated with PharmaMV to develop and tune the MPC against the mechanistic simulation of the process. The PharmaMV platform was subsequently transferred to the physical process. With this approach the MPC derived from the mechanistic model was utilized to accurately control the defined CQAs, such as final particle attributes, potency or moisture.

520705 Impact of Physico-Mechanical Properties of Co-Processed Excipients on the Tableting Performance By DM3 Approach

Abstract:

The aim of present investigation was to evaluate tableting performance of co-processed excipients (CPEs) MicroceLac® 100, StarLac® and CombiLac® at different humidity conditions using our previously published DM3 approach. The DM3 approach involves the design of experiments (D) and multivariate analysis techniques (M3) to determine the effect of material properties (M1), and manufacturing factors (M2) on product attributes like compressibility, compactibility, and tabletability. The powders were exposed at different relative humidities of 33.0 % ± 1.4 % RH, 55.0 % ± 4.8 % RH, 75.0 % ± 4.3 % RH. These exposed powders were compressed at 5mm/minute compression speed and at 100 MPa compression pressure using bench-top compaction simulator. The evaluated material properties were sorption-desorption behavior, equilibrium moisture content (%), enthalpy of dehydration and crystallization, and heat of fusion. The compression properties were modeled with 'in-die' Heckel descriptors. The material and manufacturing factors together were used to study impact on the compressibility, compactibility and tabletability of these materials. Multivariate Analysis of Variance (MANOVA), Principal Component Analysis (PCA), and Principal Component Regression (PCR) were the statistical tools used to analyze the data. Increased moisture content gave a significant reduction in TMS. On the other hand, the reduced tablet porosity indicated increased compressibility. The Heckel analysis showed a reduction in YPpl (MPa) and YPel (MPa) values indicating increased plastic deformation and elastic recovery. Reduction in WoC and WoE was also observed on increasing humidity levels. It seems that the spray dried co-processed excipients sorbed moisture on exposing at increasing humidities. The lubricating and plasticizing effect of moisture caused an increase in compressibility. However, the moisture caused weakening of interparticular bonds and decreased particle-particle interaction thereby decreasing the TMS. Thus, present investigation can be utilized for a careful selection of CPEs in the formulation.

529371 Pharmacy on Demand: On-Demand Drug Product Manufacturing in a Miniaturized and Portable System

Abstract:

Typically, in the pharmaceutical industry, drug substances (active pharmaceutical ingredients, APIs) are synthesized in one facility and formulated to finished dosage forms so-called drug products in another facility, which could be located anywhere else worldwide. Due to the complex nature of the pharmaceutical supply chain, the industry faces several major challenges when it comes to ensuring an adequate supply of quality drug products. These challenges are not only the causes of supply chain disruptions and financial loss, but can also prevent underserved and remote areas from receiving life-saving drugs. Another concern for orphan drugs that pharmaceutical company generates small sales in comparison to the development and manufacturing cost in a large footprint facility. To address all the above issues, at MIT we have developed a continuous manufacturing platform that combines both drug substance and final drug product development in a custom-built miniature system[1-3]. In this work specifically, drug product as oral tablet manufacturing from drug crystals in the miniaturized, portable, re-configurable and automated system will be presented[1,2].

The compact, portable system, roughly the size of a household oven, [72.4 cm (length) ×53.3 cm (width) ×134.6 cm (height)] is built to manufacture on-demand pharmaceutical tablets on a scale of hundreds to thousands per day. Its design combines off-the-shelf devices with custom-designed automated mechanisms. A direct compression approach is considered for making tablets. The whole unit is divided into upper process streams and lower process streams. The upper process streams start with powder feeding and end with blending. The lower process streams include powder dispensing thru to making tablets. These two process streams operate independently under high-level software control, allowing for smooth integration between the two. As this is for on-demand pharmaceutical drug manufacturing, a simplified formulation approach is adopted by eliminating excipients required for stability. Hence, only filler, glidant, lubricant, and additional functional excipients, if necessary are considered for formulation.

Several APIs having different dosage strength were investigated. Here ibuprofen (200/340 mg) and diazepam (10/250 mg) were shown as the example. Assay, content uniformity, hardness and in-vitro testing of tablets were conducted to confirm it meets U.S. Pharmacopeia standards. Formulated blended powder is freely flowing (flow function coefficient (ffc)>10) and has a good bulk density (0.65-0.72 g/cm^3) that are suitable for making tablets via direct compression. Assay of ibuprofen and diazepam tablets are found 98.8% and 103.7 % of the labeled amount, respectively. The acceptance value in content uniformity/weight variation test is 6.12 and 9.36, respectively for ibuprofen and diazepam tablets. The tensile strength of ibuprofen and diazepam tablets are 1.14 and 0.52 MPa, respectively. Tablets dissolution profiles for both of drugs meet USP standard criteria (85% dissolved in 30 minutes).

This custom-built miniature system has the potential to (1) help tackle regional drug shortages (a threatening global problem), (2) be advantageous for drugs with a short shelf life, (3) be put into immediate production of drugs based on demand, bypassing the need to stockpile drugs (e.g., for humanitarian needs), and (4) reduce formulation complexity relative to products needing yearlong stability. We envision that this system will enable the creation of small-scale, portable, fully integrated and closed loop pharmaceutical solids dosage manufacturing that can advance the treatment of human diseases at the point-of-care.

References:

1. Azad M, Osorio Caicedo J, Brancazio D, Hammersmith G, Klee D, Rapp K, Myerson A 2018. A compact, portable, re-configurable, and automated system for on-demand pharmaceutical tablet manufacturing. Int. J. Pharm. 539 (1-2):157-164.
2. Myerson A, Azad M, Hammersmith G, Brancazio D, Osorio Caicedo J. Systems and methods for the fabrication of Ttablets, including pharmaceutical tablets. U.S. Provisional Patent Application 62/621,429, Filed on January 24, 2018.
3. Adamo A, Beingessner RL, Behnam M, Chen J, Jamison TF, Jensen KF, Monbaliu J-CM, Myerson AS, Revalor EM, Snead DR 2016. On-demand continuous-flow production of pharmaceuticals in a compact, reconfigurable system. Science 352(6281):61-67.

538133 Applying Dynamic Similarity Principles to a Narrow Therapeutic Index Drug Powder Blending Process Scale-up: Laboratory Case Studies and Relevant Regulatory Experience

Abstract:

Powder blending process scale-up is one of the critical aspects to be evaluated during scientific and regulatory assessment of oral solid dosage form manufacturing. It directly impacts content uniformity (CU), especially for potent drug applications such as narrow therapeutic index (NTI) drug applications and low dose drug applications. Theoretically, geometric, kinetic, and dynamic similarity should be collectively considered during manufacturing process scale up. However, quantitative process scale-up relationships are rarely used to support powder blending processes in new and generic drug applications. In the past few years, certain powder blend process scale-up deficiencies have been identified during the scientific and regulatory reviews of the drug applications. In this work, we report our feasibility evaluation of applying dynamic similarity principles to blending process scale-up for an NTI drug powder. Data were obtained using a laboratory-scale high-shear mixer with PAT real time process monitoring. Two types of dynamic similarity rules, i.e., maintaining either Froude number (Fr) or impeller tip speed, were applied for guiding the model formulation for blending process scale. This presentation will also cover relevant regulatory review experience on powder blending process scale up. This work provides case studies which demonstrate the utility of process engineering scale-up principles in drug development and manufacturing for rational process design space selection and better product quality.

525692 Implementation of Advanced Process Control System into Continuous Pharmaceutical Manufacturing Pilot-Plant

Abstract:

Advanced process control (APC) such as model predictive control (MPC) has several advantages in compare to conventional control algorithms [1-3]. Therefore, the pharmaceutical industries are moving toward continuous manufacturing (CM) integrated with advanced process control (APC) system. This modernized integrated system open the possibilities of achieving Quality by Control (QbC) and real time release (RTR) paradigm. However, there are different levels of challenges associated with design and implementation of an efficient control system into pharmaceutical manufacturing plant. Therefore, this is still an open area of research.

The different control strategies for continuous tablet manufacturing process has been developed and implemented into our pilot-plant facility using industry standard control platform. The critical control variables are drug concertation, powder level, main and pre compression forces, tablet weight and hardness. The different control architectures under three control algorithms (PID, MPC, hybrid MPC-PID) have been implemented and compared for control of CPP's and CQA's. The flexible control system is designed in such a way that any control scheme can be easily selected. The control system is also modular in nature that can be easily transferred to any other control platform and manufacturing plant. The assurance of CQA's have been achieved through two levels strategy. In first level, the CPP's and CQA's are controlled in real time using advanced process control system while in second level, the none-confirming products are diverted in real time in waste to assure the final CQA's of qualified tablet lots. This two level strategy assures the maximum production efficiency while assuring the final product quality. The drug concentration of blend is controlled before its compaction using PAT and MPC. An efficient control strategy for powder level control in a chute placed in between blender and tablet press unit operation of continuous tablet manufacturing process has been developed, implemented and evaluated [1]. A novel noninvasive technique based on change in electric field concept has been used for real time monitoring of powder level in continuous manufacturing pilot-plant. The sensor has been integrated with a control panel (DeltaV (Emerson)) through relays and charms. The control panel then send the signal (4-20 mA) to control platform where the control strategy, a program to operate the sensor and a program to convert the sensor signal into powder level have been implemented. Subsequently, an advanced model predictive controller as well as PID controller for powder level have been implemented into our continuous pharmaceutical manufacturing pilot-plant facility. The performance of both control strategies have been practically evaluated for set point tracking and disturbance rejection. The APC has been used to control tablet press as well [2]. The pre and main compression forces have been controlled using a novel decoupled strategy so that the desired tablet weight and hardness can be achieved simultaneously. Similarly, the other CPP's and CQA's have been controlled. It has been observed that the MPC performs better than PID. A systematic methodology for material traceability has been also developed and coupled with integrated control system [3]. The methodology starts with material and process specifications followed by scanning the powder drums and transmitting the material and time information to the control platform. The control platform then utilizes the information stored in the database, generated through residence time distribution model and information collected through real time monitoring to track the material during continuous manufacturing. All the relevant data generated during continuous manufacturing has been systematically collected, stored and organized in a data hub (OSI PI) and cloud system.

The objective of this presentation is to demonstrate the performance of integrated advanced process control (APC) system implemented into continuous pharmaceutical tablet manufacturing pilot-plant.

References

1. Singh, R. (2017). A novel continuous pharmaceutical manufacturing pilot-plant: Advanced model predictive control. Pharma, Issue 28, PP 58-62.
2. Bhaskar, A., Barros, F. N., Singh, R. (2017). Development and implementation of an advanced model predictive control system into continuous pharmaceutical tablet compaction process. International Journal of Pharmaceutics, 534 (1-2), 159-178.
3. Billups, M., Singh, R. (2018). Material Traceability in Continuous Pharmaceutical Tablet Manufacturing. Pharmaceutical Technology 42 (2), 32-35, 59.

536425 The Role of Software Tools in Quality By Design: A Case Study on Monoclonal Antibody Production

Abstract:

The role of software tools in Quality by Design

A case study on monoclonal antibody production

Maria M. Papathanasiou[a], Nilay Shah[a], Efstratios N. Pistikopoulos[b]

[a]Dept. of Chemical Engineering, Centre for Process Systems Engineering (CPSE), Imperial College London SW7 2AZ, London, U.K

[b]Artie McFerrin Department of Chemical Engineering, Texas A&M University, College Station TX, 77843

Over the last few years, the U.S Food and Drug Administration (FDA) has highlighted issues with drug shortages arising mostly from deficiencies in the pharmaceutical quality and manufacturing. Moreover, regulatory bodies raise concerns that the process currently in use are not state-of-the-art, while little emphasis is given to the manufacturing process which can constitute 25% of the total product cost. Along those lines, FDA urges the pharmaceutical industry to move towards novel process setups that promise: (i) risk-free scale up, (ii) seamless monitoring and (iii) high product quality [1]. In the case of monoclonal antibodies, the aforementioned challenges come along with competition arising due to patent expiration. Based on recent reports, the global pharmaceutical industry is threatened with $194 billion loss on sales, as biosimilars are expected to take 4-10% of the market by 2020 [2]. Therefore, the necessity for the use of novel, eco-efficient processes with higher product titers becomes eminent.

In this work we present an in-silico software platform to assist computational experimentation in monoclonal antibody manufacturing. The platform comprises two main parts, namely the upstream and the downstream processing. The latter is looking into both the initial capture step, as well as the chromatographic separation unit used for the purification of the antibody. The platform is designed and executed following the systematic PAROC framework and software platform that comprises four main steps: (i) model development and validation, (ii) model reduction/approximation, (iii) controller development and (iv) controller validation under in an in-silico, 'closed-loop' fashion [3]. For the controller design we employ multi-parametric Model Predictive Control (mp-MPC) strategies that allow control laws to be easily tested and implemented on embedded device. The optimal control laws are retrievable immediately through simple function evaluations, enabling fast response without the computational requirements of a computer in the background [4].

For the upstream system, we consider a GS-NS0 cell culture system operated in fed-batch mode [5], while the downstream configuration is focusing on the simulation of the capture step as presented in [6] and the Multicolumn Countercurrent Solvent Gradient Purification (MCSGP) process [7]. Following their development, the models are used in order to evaluate the performance of the studied units under continuous operation and their performance is assessed [8]. The latter falls under the umbrella of process intensification that promises shorter production times, decreased costs, tighter product quality and significantly lower energy consumption. The presented, in-silico results indicate that continuous operation could lead to prolonged culture times and cell viability in the upstream. In the case of the downstream, the MCSGP unit is simulated under the operation of the mp-MPC controller that returns periodic input profiles and successfully handles disturbances resulting from variations in the composition of the upstream mixture. Lastly, this work focuses on the key challenges faced during the design of the aforementioned tools and how they can be addressed. Moreover, the in-silico integration of the aforementioned processes is discussed, along with suggestions to overcome issues with unavailable measurements during their experimental application.

References

[1] U.S. Food and Drug Administration, "Guidance for Industry Quality Systems Approach to Pharmaceutical CGMP Regulations," 2006.

[2] EvaluatePharma, "World Preview 2016, Outlook to 2022," 2016.

[3] E. N. Pistikopoulos, N. A. Diangelakis, R. Oberdieck, M. M. Papathanasiou, I. Nascu, and M. Sun, "PAROC - An integrated framework and software platform for the optimisation and advanced model-based control of process systems," *Chem. Eng. Sci.*, vol. 136, 2015.

[4] P. Dua, K. Kouramas, V. Dua, and E. N. Pistikopoulos, "MPC on a chip—Recent advances on the application of multi-parametric model-based control," *Comput. Chem. Eng.*, vol. 32, no. 4–5, pp. 754–765, Apr. 2008.

[5] M. M. Papathanasiou, A. L. Quiroga-Campano, F. Steinebach, M. Elviro, A. Mantalaris, and E. N. Pistikopoulos, "Advanced model-based control strategies for the intensification of upstream and downstream processing in mAb production," *Biotechnol. Prog.*, vol. 33, no. 4, 2017.

[6] F. Steinebach, M. Angarita, D. J. Karst, T. Müller-Späth, and M. Morbidelli, "Model based adaptive control of a continuous capture process for monoclonal antibodies production," *J. Chromatogr. A*, vol. 1444, pp. 50–56, Apr. 2016.

[7] T. Müller-Späth *et al.*, "Two step capture and purification of IgG2 using multicolumn countercurrent solvent gradient purification (MCSGP)," *Biotechnol. Bioeng.*, vol. 107, no. 6, pp. 974–984, Dec. 2010.

[8] K. B. Konstantinov and C. L. Cooney, "White paper on continuous bioprocessing May 20-21, 2014 continuous manufacturing symposium," *J. Pharm. Sci.*, vol. 104, no. 3, pp. 813–820, 2015.

530142 Particle Synthesis in Ultrasound-Integrated Microreactors

Abstract:

Microreactors and ultrasound are important process intensification tools with great potential especially for multiphase processes. As an example, the synergistic effect of coupling low frequency ultrasound with flow reactors to achieve cavitation have been proven successful in clogging prevention in solid-liquid flows.

In addition to their susceptibility for clogging in the presence of solid particles, microreactors are also characterized by their poor convective mixing, and the laminar velocity profile results in a wide residence time distribution which will also affect the size distribution of particles synthesized in them. To address these drawbacks, we focus on the design, modeling and experimental validation of a microreactor coupled with high frequency ultrasound. While high frequency ultrasound will not induce cavitation in the bulk liquid, the radiation force will affect particles present in the flow and deflect their trajectories. Furthermore, it can also excite secondary flow structures (known as acoustic streaming) which will improve the mixing in the bulk fluid.

In this work, we have designed an ultrasonic microreactor using a high frequency of 1 MHz, which ensures that the ultrasound is established as a standing wave in each reactor layer. Furthermore, the standing wave in the fluidic channel forms a nodal plane in its center, and the resultant acoustic radiation force will deflect all particle trajectories towards this nodal plane and thus focus the particles. This focusing effect prevents the wall deposition of synthesized particles, and in addition each particle will experience similar residence times leading to a reduced particle size distribution. The focusing efficiency of this ultrasonic microreactor is characterized using inert particles seeded with the flow. In a second step a solid forming reaction (precipitation of calcium carbonate) was performed. The effect of process parameters (e.g. ultrasound intensity, flow rate, particle concentration) on the focus extent and the particle size distribution (measured by laser diffraction) was investigated.

This work showcases how coupling high frequency ultrasound with microreactors can be exploited to improve particle synthesis.

534531 Demonstration of Scale-out Methodology for Intensified Liquid-Liquid Processes

Abstract:

DEMONSTRATION OF SCALE-OUT METHODOLOGY FOR INTENSIFIED LIQUID-LIQUID PROCESSES

*Eduardo Garciadiego Ortega, Dimitrios Tsaoulidis, Panagiota Angeli**

University College London, WC1E 7JE, London, UK

Abstract

Over the last few years there have been many studies on small scale units aiming at the intensification of multiphase processes relevant to energy and process industries. These include (bio)chemical synthesis (i.e. flow chemistry), absorption, extraction, and (nano)crystallisation. Gas-liquid and liquid-liquid processes carried out continuously in small channels exhibit fast mixing, high heat and mass transfer rates, and narrow residence time distributions when compared to conventional equipment. These advantages, in an industrial context, would translate to reduced costs, energy savings, diminished waste production, enhanced safety, and increased product quality.

The transition of these novel and efficient devices from bench-scale to the industrial application requires the throughput to increase without losing the merits of the small scales. The most common strategy to address this is by reproducing the small scale units in parallel in order to have the required throughput; this approach is termed scale-out or numbering-up. It contrasts the traditional scale-up where the size of one unit increases. Flow distributors are necessary to parallelise a process. Their effectiveness is critical to bridge the gap between bench and industrial scales for intensified processing. This challenge is accentuated in multiphase processes where maldistribution in either the residence time or the phase ratio in the main channels negatively affects the performance.

In this work double manifolds are considered for the two-phase flows distribution (Figure 1). They have several advantages, including, simple arrangement, small footprint and modularity. To model the flow distribution in the double manifolds, a new model has been developed following the resistance networks approach, similar to Al-Rawashdeh's model [1]. The model can be used to select the sizes in every section of the double manifold for a given maldistribution tolerance and required throughput.

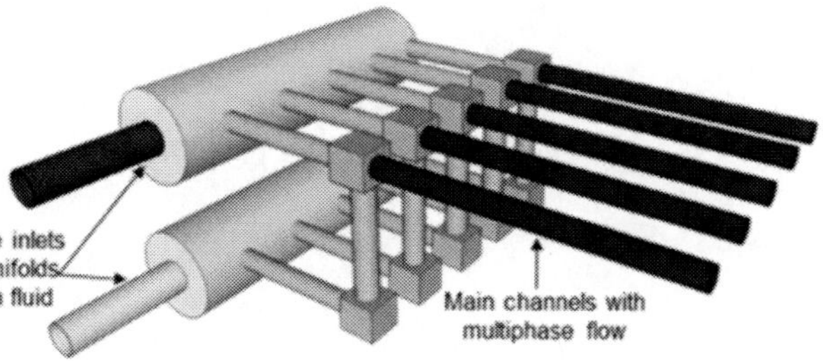

Figure 1. Schematic of a double manifold. Each fluid has a separate inlet; then both phases are distributed independently and mixed together into the main channels.

We present a demonstration of a step-by-step methodology to scale-out intensified liquid-liquid extractions. The process studied is the continuous extraction of U(VI) from a nitric acid solution into an organic phase using TBP as extractant. Previous to the scale-out, the single-channel process is characterised for conditions which result in plug flow. Hydrodynamic characteristics, such as features of the flow pattern (e.g. plug length and film thickness) and pressure drop, are related to the mass transfer performance. The results from the single-channel process are used to design a double manifold for the numbering-up. The resistance network model is used to decide the channel sizes in the manifold. Extraction studies are then carried out in the scale out system.

References

1. M. Al-Rawashdeh, F. Yu, T.A. Nijhuis, E. V. Rebrov, V. Hessel, J.C. Schouten, "Numbered-up gas-liquid micro/milli channels reactor with modular flow distributor," Chem. Eng. J. 207–208 (2012) 645–655. doi:10.1016/j.cej.2012.07.028.

537806 Multiphase Microchannel Separation Utilizing Capillary Pressure Gradients

Abstract:

A flexible, modular platform for multiphase separation is being developed for downstream processing that shows potential for significant reduction of capital cost and energy expenditure compared to conventional technology. The separator utilizes microchannel processing technology (MPT) to direct flow of each phase by creating a capillary force gradient via size and spacing of micro-scale architectural features, thereby controlling interfacial curvatures. With a proper selection of surface properties, the system is designed so that a selected phase cannot overcome capillary forces (in one direction of the gradient) with inertial and viscous forces of flow, guiding the fluid towards one outlet stream. The separator uses a flat â€œLamina Plateâ€ design that can accommodate a larger processing throughput per layer of device compared to single microchannel configurations. Microchannel-based separation presents significant gains in process intensification by accelerating mass transport and decreasing energy input per liquid volume compared to conventional methods. Beyond energy consumption, the scalable, distributed nature of MPT makes it amenable to distributed, smaller chemical manufacturing.

529629 Effect of Pressure on Ethane Dehydrogenation in MFI Zeolite Membrane Reactor

Abstract:

Using a membrane reactor (MR) for producing ethylene by ethane dehydrogenation (EDH) reaction is an effective way. Compared with packed bed reactors (PBR), the EDH MR effectively surpasses the equilibrium limit by timely removing H_2. A packed-bed membrane reactor (PBMR) with a Pt/Al_2O_3 catalyst was used to investigate the effect of pressure on EDH reaction. The EDH reaction was performed in the PBMR for the pressure and temperature range of 1-5 atm and 500-600 ^oC, respectively. With an increase in reaction temperature, the reaction rate increased which caused higher ethane conversion. Increasing reaction pressure helped in enhancing H_2 permeation across the membrane, which significantly increased the ethane conversion. The equilibrium limit of ethane conversion was successfully surpassed by increasing temperature and reaction pressure in the PBMR. Ethane conversion and ethylene selectivity as high as 29% and 97% were obtained at 600 ^oC and 5 atm for PBMR while corresponding values were 7% and 75% for PBR. The timely removal of H_2 from the reaction side also helped in reducing methane formation as H_2 is required for the methanation to occur. In addition, a 1D plug flow model was developed and the values for ethane conversion obtained from the model were validated with experimental results. The same model was used to evaluate the ethane conversion beyond the experimental conditions, showing ethane conversion >90% could be obtained.

537204 Scaling a Micro Structured Reactor for Sugar Chemistry from Lab Via Pilot to Full Production Scale

Abstract:

Scaling a micro structured reactor for sugar chemistry from lab via pilot to full production scale

M. Kraut[1)], G. Rabsch[1)] and R. Dittmeyer[1)]

1) Karlsruhe Institute of Technology, Institute of Micro Process Engineering, Hermann von Helmholtz Platz 1 76344 Eggenstein-Leopoldshafen, Germany

Motivation

It is well known, that micro structured devices show very high heat transfer rates, and some processes profit immensely by fast heating and cooling processes. Examples of production processes performed in micro structured reactors are still quiet rare. Following Roberge's work [1] it is surprising that not more processes have been transferred from macro scaled devices to micro or milli-scaled ones.

Especially processes in the food industry are often considered not to be compatible with micro structured flow devices due to the properties of natural products (varying quality, unknown side products, fouling tendency…) The challenge is taken up to make a new product using micro structured heat exchangers to obtain high specific heating rates to achieve defined residence times to perform a sugar transformation.

Process

The transformation of sugars at elevated temperatures is well known for a long time (e.g. caramel making). The target product is narrowly defined in terms of selectivity especially of undesired side products. This quality is achieved by a very well defined residence time of the reaction mixture at an equally well defined temperature. The reaction mixture firstly is brought to a temperature which is safely below the onset of both, desired and undesired reactions. Then the mixture is heated up rapidly (heating rates >100 K/s) to the optimum reaction temperature and held there for a defined time to obtain equilibrium yield of the desired and a minimum yield of undesired side product. After water is separated from the reaction mixture as steam the procedure is repeated with a second set of temperature and residence time.

Laboratory reactor

In Lab-scale the task was to find the parameters for operating the process reproducibly and safely within the required quality criteria. As a tool for preheating a symmetrical cubic heat exchanger was applied, which was described previously [2]. To perform the reaction and provide the necessary residence time a micro structured so called laboratory module[3] was used (fig 1, bottom).

This module provides 30 parallel reaction channels and for an intense heating 179 channels perpendicular to the reaction channels. The structured foil has been made from stainless steel (1.4571 or 316 Ti) and structured mechanically. The reactor is operated with a very high volume flow of thermos oil to obtain the boundary condition of a "hot wall". Since the temperature difference in the oil is negligible a heat balance can't be done, but the point is to provide the reaction with optimum conditions, not the most efficient heat usage.

Since the reaction itself has a very small enthalpy the heating effort is small, and one could argue that the reactor is over-designed. However, even small variation in temperature can lead to undesired effects, so that the condition to keep the temperature constant is sufficient to design a completely temperated reactor.

The process parameters for the reaction were found to be in the range of a few dozen seconds in residence time and a reaction temperature of about 200°C.

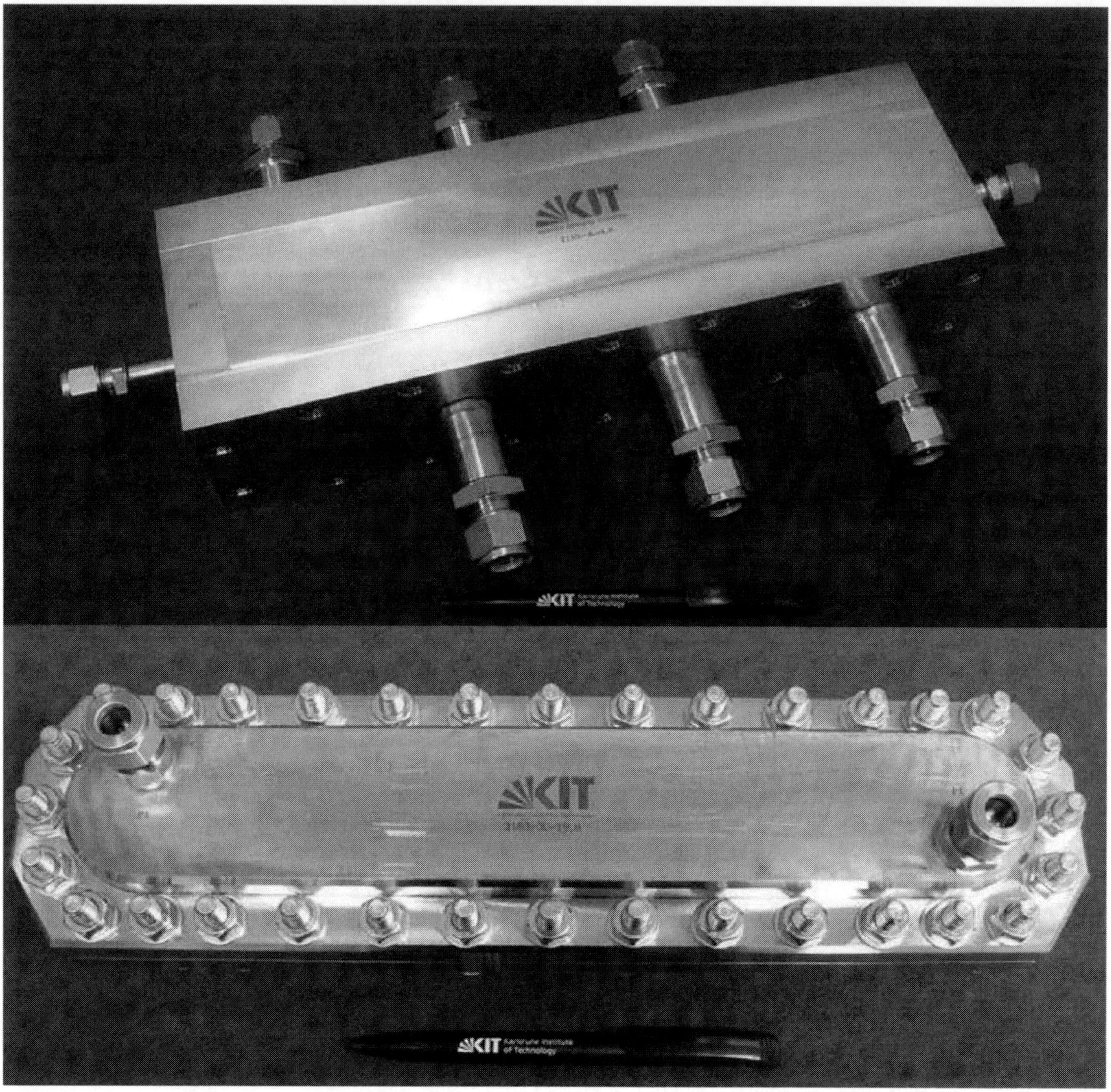

Fig.1 KIT lab module (bottom) and Pilot reactor derived from lab module (top)

Pilot scale reactors

The lab results were then transferred to a pilot scale. The scaling principle was to keep the structure of the device and multiply it according to the scaling factor, in this instance a factor of 15 (from 1 to 15 kg/h). Since the parameters of operation were known, the pilot device was manufactured in several layers of plates for heating and cooling respectively. The thickness of each foil was 1 mm, so the foils were diffusion welded to form a monolithic block (fig.2). The length of the heating channels was slightly increased to ensure diffusion welding compatibility of the device. However, geometry for the inflow has been altered, so that the specific pressure drop was not increased in comparison to the laboratory module. Also a 16th layer of heating channels was added to obtain the same conditions for each of the reaction channels (heated walls in both directions).

The reaction conditions were successfully reproduced at designed thoughput using this device. No detrimental scaling effects were observed. However, the flow distribution in the flanges was optimized during the pilot phase to ensure a better distribution based on cfd-calculations. [4]

To test the possibility of etching as an alternative manufacturing technique, which for higher amounts of foils is economically advantageous, one device was manufactured using etched foils. Dimensions were slightly altered to obtain circular channels by face-to face stacking. That device was also tested successfully.

Fig.2 Diffusion welded block of pilot reactor. 16 layers of cooling foils are visible. Three cooling zones are evident.

Production reactor

Based on these encouraging results the design of a production unit with a further scaling factor of 60 was done. Since a reactor with a factor of 60 in height which would be the consequent scaling option was found to be not feasible; several conditions were altered:

- Crossflow heat exchange was replaced by countercurrent heat exchange
- The number of parallel reaction channels was increased
- The heating fluid is divided in two streams, each of which is flowing through a half of the reactor.

Design calculations ensure the condition of "hot wall" or constant reaction temperature along the whole channel.

Fig.3: Design of production reactor as scaled up version of the pilot device, reaction direction ion green, heating in red.

Conclusions

A process of sugar transformation has been investigated using micro structured reactors and found to be feasible when using well defined residence time at equally well defined temperature, exploiting the high heating rates and flow conditions within micro structured channels.

A scale-up from laboratory device to a diffusion welded pilot device has been done successfully, showing that the scale-up strategy of using identical local conditions is feasible.

Further scale-up to obtain a micro structured reaction system has been performed with further economic improvements.

[1] Roberge, D.M., Ducry, L., Bieler, N., Cretton, P., Zimmermann, B., Chemical Engineering and Technology 28(2005), pp. 318-323

[2] Schubert, K., Brandner, J., Fichtner, M., (...), Schygulla, U., Wenka, A. , Microscale Thermophysical

Engineering5(2001), pp. 17-39

[3] K. Haas-Santo, M. Kraut, W. Benzinger, A. Wenka, L. Bohn, R. Dittmeyer, Chem. Eng. Tech., 82 (2010), p. 1341

[4] J. Cao, M. Kraut, R. Dittmeyer, L. Zhang, H. Xu, Int. Commun. Heat Mass 93 (2018), 60-65

521591 Green Synthesis of Polyvinyl Butyral (PVB) Using Microreactor System and Recycling Technology

Abstract:

We propose a highly efficient method for the synthesis of polyvinyl butyral (PVB) using microreactor system. The condensation reaction between polyvinyl alcohol (PVA) and n-butanal was successfully implemented, allowing the energy costs associated with cooling and mixing of the reactant solutions to be reduced. The microreactor system allowed the development of a recycling technology for the reuse of HCl, water, and excess n-butanal present in the product solution. In addition, with the aid of a two-step n-butanal feeding method, the developed recycling technology resulted in a 98.7% n-butanal utilization ratio, while the consumption of HCl and water during the reaction process was reduced by 85.6%. These results indicate that our novel process represents a more environmentally friendly approach to the PVB synthesis.

Funding support from National Key R&D Program of China (2017YFB0307102) and Tsinghua University Initiative Scientific Research Programs (20131089196, and 20151080361).

537752 Client-Sponsored and Team-Defined Senior Capstone Projects

Abstract:

The Departmentâ€™s capstone experience allows students to apply their chemical engineering education to solve an open-ended design problem. In the past, teams have applied the engineering design process to paper-based projects, such as the AICHE-sponsored design challenge. In recent years, we have collaborated with local industrial sponsors seeking to optimize part of a manufacturing process. Students must coordinate site visits with the client to observe the process under review and work closely with the industrial partner to deliver expected results. Additionally, student teams are required to perform hands-on experimental trials in the Unit Operations laboratory to test specific design hypotheses. Recent project examples include: 1) Assess efficiency/improvement opportunities for manual assembly line at the B.Braun, Dominican Republic site. 2) Provide a preliminary process design for the recovery of edible protein from cheese whey at Klein Farms. 3) Assess efficiency/improvement opportunities for the watercolor assembly line at Crayola. 4) Assess efficiency/improvement opportunities for bottle filling line at FMI Cosmetics.

This year, the Department expanded the capstone project offerings to include a Team-Defined project. These student-proposed design projects are comparable in scope to client-sponsored projects and can be extensions of student summer internships. Team-Defined projects also present a great opportunity to develop an entrepreneurial idea of interest to members of the team. In the case of an entrepreneurial idea, there is an added requirement of a more detailed pro forma for the proposed business. Team-defined project examples include: 1) Provide a preliminary process design for a commercial plant for 1) kombucha production, 2) apple cider vinegar production, and 3) orange juice concentrate production. Student teams are required to perform hands-on experimental trials in the Unit Operations laboratory to test specific design hypotheses, and also simulate the commercial scale plant in ASPEN.

In this presentation, several projects will be highlighted to demonstrate achievement of course objectives, which are: presenting technical material to an audience at an appropriate level of difficulty and approach; working in teams to effectively and ethically overcome engineering problems and technical challenges; demonstrating project management, organization, and documentation skills for the effective operation of an engineering development team; performing an economically sensitive process or product design project to demonstrate an understanding of cost, profit, regulation, and market issues; and preparing an exhaustive engineering technical report that relates a project in such a manner as to satisfy the goals of clarity and completeness in detail and presentation.

In our experience, client-sponsored and team-defined projects have been well received by students looking for a practical application of their education in their final capstone experience, and present an acceptable alternative to paper-based design.

524225 Embedding Engineering Research in Process Design Education

Abstract:

Technical literature search is undoubtedly one of the most important components of chemical process design. However, many undergraduate students graduate without knowing how to effectively search, synthesize, and utilize knowledge from vetted technical literature (e.g journals, patents, standards, handbooks) in engineering design settings. Students tend to over rely on surface-level Google searches, while underutlizing powerful information discovery toolsets that are available to them through organizational subscriptions, such as Knovel and Engineering Village. Research skillsets are most successfully learned in context of an actual project, rather than a one-off lecture. To this aim, Research Sprints are a fast-paced student-centered learning experience that combine elements of design thinking, agile processes, and engineering research together, to layer in technical acumen with collaboration and communication skillsets.
Knovel is an engineering decision support tool that builds upon a rich corpus of engineering textbook and material property content to serve engineering analytics use cases. Engineering Village complements Knovel with deep indexing of engineering journal, conference, IP, standards, and funding literature. In this talk, we will demonstrate how Knovel & Engineering Village can be used together into the classroom to expose students to technical literature and apply the insights in engineering design.

533875 Effective PBL in Senior Design: Examples & Lessons

Abstract:

With problem-based learning (PBL) students learn by solving an open-ended problem. Education research suggests that PBL is one of the trickiest instruction modes to implement. For example, some studies found that PBL success is more sensitive, compared to other instruction methods, to the subject matter 'expert level' of the instructor. PBL furthermore tends to fail more drastically when it becomes overly student self-directed (Prince, Journal of Engineering Education, 93(3), 2004). Virtually every BS Chemical Engineering program in the USA culminates with a capstone design module that is inherently a PBL exercise. Is this experience always, or even typically, as effective a learning experience as it could be?

This presentation will share a few examples drawn from two new BS Chemical Engineering programs, the first at the Rochester Institute of Technology (first graduating class 2013) and the second from the University of Iceland (first class 2017). New programs afford a unique 'blank sheet' opportunity to assemble a complete curriculum from scratch with no historical baggage and test alternative approaches.

The presentation will include a brief survey of design sequence curriculum choices across institutions in the USA. Selected examples of tried-and-tested design problems with specific learning outcomes will be shared. The importance of curriculum integration will be illustrated.

537284 Incorporating Laboratory Pilot Plant Data and Customized Software Simulations into Chemical Engineering Design Education at Vanderbilt

Abstract:

Russell F. Dunn, Ph.D., P.E.

Professor of the Practice

Department of Chemical and Biomolecular Engineering

Vanderbilt University

Nashville, TN 37235-1604

Ph: 615-343-4268

russell.dunn@vanderbilt.edu

Matt J. Lang, Ph.D.

Professor

Department of Chemical and Biomolecular Engineering

Vanderbilt University

Nashville, TN 37235-1604

Ph: 615-345-7493

matt.lang@vanderbilt.edu

Bryan R. Beyer

Adjunct Instructor

Department of Chemical and Biomolecular Engineering

Vanderbilt University

Nashville, TN 37235-1604

Ph: 615-343-4268

bryan.r.beyer@vanderbilt.edu

Incorporating Laboratory Pilot Plant Data and Customized Software Simulations into Chemical Engineering Design Education at Vanderbilt

Integrating unit operations laboratories with scale-up design assignments offers unique perspectives for the students to connect equipment pilot equipment operation to industrial scale design. Two-week laboratory modules involving reverse osmosis and distillation technology, respectively, are currently run by undergraduates students at Vanderbilt as part of their senior unit operations laboratory course. Students then use this pilot scale data to generate a traditional laboratory report covering these respective technologies. An additional week is then spent by the students using customized water network design software is coupled with the reverse osmosis technology to design an industrial scale water interception and recycle network. Likewise, customized heat exchange network software and Aspen simulations are coupled with the distillation technology to design an industrial scale distillation column with heat integration. Two examples illustrating these approaches will be shown.

534970 Implementation of Joint Senior and First-Year Design Projects

Abstract:

Students in their final year of chemical engineering complete a design project that requires them to use the material learned over the course of their studies. In their undergraduate curriculum, the students acquire design and project management experience but very little leadership experience. At the same time, incoming first-year chemical engineering students have a difficult time connecting their introductory mathematics and science courses to their future engineering courses. While freshman students take an introduction to chemical engineering course with a design project, they lack the technical knowledge and skills to do meaningful projects. To help with all of these problems, we created a joint design project where first year students work as interns on a senior design team project that focuses on product design. The two classes share a one hour meeting time each week, as well as having separate class time. The senior students act as supervisors and mentors for the first-year students and give assignments and provide feedback for the first-year students. The design product is divided into four phases: (1) product idea development (2) final product selection (3) process design, and (4) final design. In addition to the assignments from the senior students (emphasis in phase 1 and 4), first-year students receive overarching instructor given assignments within each phase of the project. For phase 1, both groups conduct market research to determine product opportunity gaps. The first-year students are required to submit a report detailing their research and market analysis, but the specific research assignments were given by the mentor teams. Additionally, the first-year reports were submitted to the senior teams for feedback prior to being submitted to the instructor. For phase 4, each member of the first-year team becomes an expert on one of the following areas: safety, economic impact, or environmental impact. The student then researches and reports on this topic to both the instructor and the mentor teams. In phase 2, the first-year teams work with the mentor teams to create a weighted selection matrix and narrow down product ideas, culminating in a sales pitch, advocating for one particular product selection, to the mentor teams. Mentor teams select the final design idea. Phase 3 will have the most separation between the teams with the senior teams creating the actual product process design, and the first-year teams creating a product in a hands-on laboratory, a simplified block diagram, and a material balance for the process. In the first two years, both senior and first-year teams were surveyed at the end of the semester. This paper focuses on the results of implementation of the project in its first two years, and changes made for the second year implementation based on feedback from the first year.

540314 Maintaining Operation Excellence in Badak LNG in Facing Leaner Gas Entrance

Abstract:

Badak LNG as one of the pioneer in LNG business is currently facing operational challenge in operating plant safely, efficient, and reliable. One of the current operational issues faced in Badak LNG plant is the leaner gas entrance to the facility. The leaner gas composition may cause inefficient operation as it requires some amount of LPG being reinjected to the feed gas stream to maintain a desired LNG quality specification.

The Badak LNG plant facility is designed with specific feed gas quality (±83% of Methane content) and single LNG quality. Along time, leaner gas is introduced to the feed gas which has caused the Methane content to reach 90%. This condition may cause the LNG quality (i.e. heating value/HHV) to drop below the Rich LNG sales contract specification. As countermeasure, some amount of LPG was reinjected to the feed gas stream in Badak LNG facility to maintain the LNG product quality as per specified quality.

With the shifting operation mode, Badak LNG is challenged to optimize this by manipulating the operation mode inside the plant. Badak LNG has implemented an operational breakthrough by segregating LNG production line of production. This was done by separating Process Train, LNG rundown line, LNG Tanks, and LNG Transfer Line, & Terminal to produce 2 LNG qualities in parallel. This method was later known as dual HHV mode. A single production line is set to produce rich LNG (HHV > 1,107 BTU/SCF) to fulfill rich LNG sales contract, while the other production line is set to produce lean LNG (HHV < 1,107 BTU/SCF) for the other sales contracts or spot markets. With this operation, the LPG consumption can be reduced significantly with a potential saving of almost 300 kUSD/day.

534198 Evaluation of CO_2 Capture and Storage Systems for Existing Thermoelectric Plants in Mexico

Abstract:

Global warming has increased in recent years due to greenhouse gas emissions derived from all human activities, being the power generation from fossil fuels the largest source of CO2 emissions. In Mexico, according to the Federal Electricity Commission (CFE acronym in Spanish), in 2012 there was an installed capacity for electricity generation of 53.114 GW, from which approximately 72.6% of the total electric power generation came from thermoelectric power plants, whether conventional, combined cycle or internal combustion.

CO2 capture and storage has been recognized as a safe and environmentally acceptable solution for continue using fossil fuels with minimal CO2 emissions. Among the existing capture technologies, post-combustion capture by chemical absorption is the most popular, as it may be retrofitted to existing power plants. The main objective of this work was to determine the feasibility of their implementation in Mexico, bearing in mind the installed thermoelectric capacity. The required potential for CO2 capture, as well as the energy and solvent demand during the capture processes were estimated for each existing power plant. All the estimations are computed using models form the literature, considering the different fuels and modes of energy production.

Furthermore, as the Mexican economy is largely based on oil production, the CO2 recovered from power plants can be used for improving the oil recovery from near-exhausted wells. Emissions from the stationary sources can be reduced by capturing CO2 and then sequestering it from the atmosphere through geological storage. Many oil and gas reservoirs are excellent candidates for geological storage, since CO2 can be injected into these reservoirs to provide additional pressure used for the oil recovery, due to CO2 reduces the oil viscosity and improving its displacement to the production well. Capturing CO2 from the selected sources and storing in the suitable sites are critical for developing a cost-effective process. In view of the estimated capture capacity, linear optimization models are formulated to determine the best distribution paths, considering the energy expenditure in direct CO2 transportation from power plants to sequestration sites, as well as the existence of intermediate distribution hubs.

534225 Application of Ion Selective Electrodes (ISEs) for Real Time Monitoring of Flue Gas Desulfurization (FGD) Wastewater

Abstract:

Coal-fired power plant produces a significant amount of SO_2 as well as heavy metals during operation. Flue Gas Desulphurization (FGD) plays a very important role in this regard to reduce the emission of SO_2 and heavy metals into the environment. However, FGD process produces wastewater during the process which contains these harmful elements. So, it is necessary to treat this wastewater and observe the concentration of targeted species according to the EPA guidelines before releasing in the environment.

Ion Selective Electrode (ISE) can act as an important tool to continuously monitor the wastewater discharge for targeted species. ISE is considered as a sensor that can detect targeted species by converting the movement of those species in a dissolved solution into an electric potential. However, to implement ISEs for wastewater monitoring, it is necessary to evaluate their features over a large range of conditions.

This aim of this study was to analyze the characteristics of several ISEs that can be utilized in FGD wastewater treatment plant. Calcium, chloride, nitrate, and nitrite ISEs were calibrated in a broad range of concentrations. Multi-parameter regressions found that each ISE fits well over a wide range of concentrations with various temperatures and ionic strength interactions. Based on the laboratory experimental data a robust sensor module with multiple ISEs will be built, calibrated, and tested at a real coal fired power plant.

533717 Energy Efficiency Calculation of a Combined Heat and Power(CHP) Plant Integrated with Torrefaction Process Using Aspen PLUS

Abstract:

According to the climate policy in the European Union, we should reduce the greenhouse gas emission and increase the portion of renewable energy in the final energy. So, biomass is recognized as an important renewable energy source and contributes to approximately 10% of the global energy consumption. Combustion of biomass is almost used for heat generation in the small scale plants, or combined heat and power (CHP) plant. Torrified biomass shows similar characteristics those of coal, like comparable heating value, hydrophobicity, good grindability and so on. Therefore, Torrified biomass can be co-combusted with coal.

Torrefaction is a pretreatment method for improving biomassâ€™s thermal property as a fuel. The procedures of torrefaction process is as follows : drying, torrefaction, cooling. The process is simulated in ASPEN PLUS V8.8. A drier using air is simulated in our model. Through the drier unit, the feedstock biomass moisture contents is reduced from 50wt% to 10 wt%, wet basis. Torrefier is the main reactor of this process, where the biomass feedstock is thermally treated to produce torrefied biomass and by-products. Due to the complexity of torrefaction, it is difficult to use a pre-defined reactor in Aspen Plus to model this process. Consequently, it normally requires several blocks and calculators with FORTRAN codes to simulate this reactor. Reduction of mass and energy yield and improvement in heating value of the torrefied biomass are observed and optimal temperature of torrefaction are between 275 and 278 degrees. Total Energy consumption and process energy efficiency is calculated from the model. The drier consumes 76-80% of the total heat demand.

A CHP plant model is also simulated in ASPEN PLUS. There are three part of the process : fuel combustion, steam generation, steam turbine cycle. In fuel combustion part, we use Yield reactor and Gibbs reactor to simulate fuel combustion with calculator box. To make superheated steam, the boiling feed water pass an economizer, evaporator and superheater. Superheated steam is led to a High pressure(HP) turbine and the outlet stream is separated back to boiler to reheat the used steam. The reheated steam is injected Intermediate pressure(IP) turbine with HP turbine outlet stream. Finally the steam pass through the low pressure(LP) turbine and condenser. The condensed water is heated by the steam extracted from each turbine, pass boiling feed water pump, recycled to the economizer. From the CHP model, we can calculate electric efficiency and total energy efficiency. Conventional CHP plantâ€™s electric efficiency is about 30 to 40% and total energy efficiency about 70 to 80%. And will perform a case study to replace 25%, 50%, 75%, 100% of the fossil fuel in boiler using the torrefied biomass. Compared with coal, torrefied biomass have competitive both efficiencies on the model and reduce the emission of CO_2.

Within the CHP plant using steam turbine, the possible spots of integration of torrefaction in CHP plant are various because of operating temperature of torrefaction including drier, usually 100 to 300 degrees. In flue gas from the combustion, we can easily find the integrating points due to stability of the flue gas temperature. So, we devide the scenarios of integration: the flow after HP turbine, IP turbine and flue gas stream. After this, we calculate energy efficiency of the integrated process, and determine whether the torrefied biomass can replace fossil fuel 100%.

536124 Utility System Optimization Under Air Quality Considerations

Abstract:

Air pollution is a primary concern, and one of its major contributor is industry. Optimization of emissions from industrial facilities is well established in publications and reports. However, it does not extend beyond the stack or exhaust level i.e. emissions released to the atmosphere. This work aspires to develop and test a methodology to optimize the operational scheme of a utility system considering air quality in the area around the facility by combining Process Design (PD) with Atmospheric Dispersion Modelling (ADM). The concept of this work is to investigate if ADM can be used to improve pollution prevention/control by observing ground level concentration in the surrounding area of the unit at various operating scenarios and in different weather conditions. The methodology has been implemented in MATLAB following the coupling of a simple Gaussian dispersion model with a process model, both supplied with real meteorological and process data. The case study used to test the methodology is a High Pressure Steam (HPS) generation unit. It consists of three identical boilers where the operational strategy is to operate two boilers, and the third is stand-by. The boilers accept two types of fuels: gas and liquid. Two separate optimization goals were studied. The first is to improve air quality in the surrounding area, i.e. minimize the ground level concentration, by changing the operating scheme of the three boilers. The second optimization goal is to reduce the operating cost by optimizing the ratio of the two fuels while keeping the resulting ground level concentration just below the regulations limit.

Finally, the methodology allowed to derive and assess the different operational strategies. The first goal improved the overall air quality and reached up to 46% reduction of the maximum concentration exceedances. The second goal proved that facilities can reduce operational cost and still be in compliance with environmental regulation. On the other hand, this cost reduction does lead to a decrease of overall air quality since average ground level concentration is increased. In the future, this methodology could be applied to more case studies and at industrial city level to improve air quality and to assist on more appropriate environmental policies.

525883 Design of Extractive Distillation Processes Using Simulated Annealing and a Rigorous Process Simulator

Abstract:

Extractive distillation is a widely-used method for the separation of azeotropic mixtures, and many researchers have reported on the design and optimization of extractive distillation processes [1â€"7]. Due to the large number of design variables, most optimization work is either based on simplified column models, or is conductive iteratively one variable at a time. The former method is less accurate due to the simplified column model and the latter method is susceptible to being trapped in a local minimum.

In this work, continuous extractive distillation processes were optimized using simulated annealing (SA) and simulation in a rigorous process simulator (Aspen Plus). Connection between the two programs was achieved by means of a client-server architecture, with Matlab as the automation client and Aspen Plus as the automation server. Three separation problems were considered: separation of acetone and methanol, separation of hexane and ethyl acetate and the separation of n-hexane and tetrahydrofuran. For the separation of acetone and methanol, previous results for the process design are available in the literature for comparison. For the latter two separations, to our knowledge there are no reports of continuous process designs for the separation in the open literature. For each separation, several different candidate entrainers were considered and the costs were compared.

For the separation of acetone and methanol, the TAC of the process for each entrainer considered was lower than the initial condition taken from the literature. For the separation of n-hexane and ethyl acetate, NMP, 2-methylpyridine, 3-methylpyridine, DMF, and pyrrole were considered as possible entrainers. The results show that DMF offers the best performance and economic benefit. For the separation of n-hexane and tetrahydrofuran, the candidate entrainers were DMF, NMP, and 2-methylpyridine. DMF was again found to be the most suitable entrainer. The results show that the SA algorithm has the advantage of running automatically and has a high probability to obtain a design near the global optimum.

[1] Luyben, W. L., & Chien, I. L. (2011). Design and control of distillation systems for separating azeotropes. John Wiley & Sons.

[2] Widagdo, S., & Seider, W. D. (1996). Journal review. Azeotropic distillation. AIChE Journal, 42(1), 96-130.

[3] Andersen H W, Laroche L, Morari M. Effect of design on the operation of homogeneous azeotropic distillation. Computers & chemical engineering, 1995, 19(1): 105-122.

[4] Knapp, J. P., & Doherty, M. F. (1990). Thermal integration of homogeneous azeotropic distillation sequences. AIChE journal, 36(7), 969-984.

[5] Kossack, S., Kraemer, K., Gani, R., & Marquardt, W. (2008). A systematic synthesis framework for extractive distillation processes. chemical engineering research and design, 86(7), 781-792.

[6] Kossack, S., Kraemer, K., & Marquardt, W. (2006). Combining shortcut methods and rigorous MINLP optimization for the design of distillation processes for homogeneous azeotropic mixtures. In INSTITUTION OF CHEMICAL ENGINEERS SYMPOSIUM SERIES (Vol. 152, p. 122). Institution of Chemical Engineers; 1999.

[7] Garcia-Herreros, P. and J. M. Gomez (2011). "Optimization of the Design and Operation of an Extractive Distillation System for the Production of Fuel Grade Ethanol Using Glycerol as Entrainer." Industrial & Engineering Chemistry Research 50(7): 3977-3985.

531755 Comparison of Batch Vs Continuous Operation for the Acquisition of Robust Reaction Kinetics in the Production of Succinimide Dispersants

Abstract:

Production of specialty chemicals to-date is conducted almost exclusively in batch or semi-batch reactors. Transition to continuous processing promises strong improvements—so-called "process intensification"—over the century-old batch processing technology with lower unit energy consumption, less waste generation, reduced exposure hazards, smaller process footprint, lower capital investment and more consistent product quality. The transition to continuous processing—already demonstrated widely at the much smaller scale of the pharmaceutical industry—thus offers the prospect of truly revolutionizing the manufacture of specialty chemicals. However, the detailed chemistry and kinetics underlying current processes are in most cases poorly understood (if not completely unknown). Transition to continuous processing hence requires collection of robust and accurate kinetics as a key prerequisite.

This work presents results from a university-industry collaboration between Lubrizol Corporation and the University of Pittsburgh, under the umbrella of the recently formed RAPID Modular Chemical Process Intensification Institute. It aims to demonstrate the transition from batch to continuous processing focused on lubricant chemistry. Specifically, we present results from a kinetic investigation into the production of succinimide dispersants via reaction of a poly-isobutylene succinimide anhydride (PIBSA) with polyamines. The molecules undergo a condensation reaction to produce a large surfactant molecule with a polar head group and a long, oil soluble polymer tail. A complete family of products (which form the largest volume oil additives) is made by varying molecular weight, structure and stoichiometric ratio of the various reactants.

Here, we present results from a detailed investigation of the reaction system in a small laboratory-scale batch reactor using online IR spectroscopy as the analytical tool. The use of a batch system has the advantage of easy experimentation and enables the acquisition of large amounts of data through measurement of long concentration time traces, thus facilitating the acquisition of robust kinetics. However, in order to cover a large parameter range (such as operating temperatures, reactant concentrations, etc) a large number of experiments is required. We therefore compare the acquisition of kinetic data from the batch system with a continuous reactor system and discuss advantages/drawbacks of each variant.

533977 Implementation of a Control Strategy for a Multitasking Reactive Distillation Column with an Intermediate Condenser

Abstract:

With the increasing energy demand around the world, global warming, and the continual decrease in fossil fuel reserves, there is growing interest in many countries in renewable, clean and sustainable energies. Most photovoltaic solar cells use various crystalline and amorphous structures of silicon as their semiconductor element. These, can be obtained from silane as a precursor. Silane is commonly synthesized as a result of the disproportionation reactions of trichlorosilane, in a process involving two reactors and various separation units. However, the unfavorable chemical equilibrium in the reaction system causes a substantial recycle ratio, with consequential high investment and energy costs. As a result, previous studies have proposed and designed a new reactive distillation column (RD). This takes advantage of the chemical characteristics of the disproportionation reactions and the significantly different boiling points of their components. The incorporation of an intermediate condenser within the rectification zone in the reactive distillation may alleviate the refrigeration load of the overhead condenser. The optimal design and operating conditions for this RD with an intermediate condenser have been evaluated under steady state conditions. This paper studies the dynamic behavior of the formerly explained system, working in a multitask mode to produce silane, monochlorosilane, and dichlorosilane, with high purity, by adequately switching the operating conditions. Three different control strategies were tested against perturbations in the intermediate condenserâ€™s refrigeration duty, and the associated side stream. The results showed that the best configuration implies a dual temperature control. The reactive distillation column with an intermediate condenser was capable producing the three products, with a purity equal to, or higher than, 99.5%. This control strategy allows successfully stabilization of the system after the tested perturbations. Additionally, a proposal for controlling the rectification zone temperature with the refrigeration load of the intermediate condenser, was also built and tested.

536482 Aqueous Benzyl Alcohol Oxidation Using Polymer Nanoreactors: Towards Multifunctional Nanoscale Reactors

Abstract:

Confining catalyzed reactions to the hydrophobic core of micellar nanoreactors can facilitate organic phase reactions in a bulk aqueous environment. We are investigating confining catalyzed reactions to the hydrophobic core of kinetically-trapped polymer nanoreactor to achieve reaction and separation by leveraging the biphasic nature of the reaction. Core-shell nanoreactors encapsulating catalytic gold nanoparticles have been fabricated through block copolymer directed self-assembly. The focus of this work was evaluating use of the nanoreactors for oxidation reactions using conversion of benzyl alcohol to benzaldehyde as a model reaction. Using polystyrene as the microenvironment two oxidants were investigated: air and hydrogen peroxide. Use of hydrogen peroxide as an oxidant resulted in a 10-fold increase in benzyl alcohol conversion compared air. We also investigated the effect of hydrophobic microenvironments on catalytic oxidation using polystyrene, vitamin E, and perflourodecalin. The presence of Vitamin E in the nanoreactor core inhibited the reaction due to its anti-oxidant properties. The presence of perfluorodecalin nanoreactors increased the conversion of benzyl alcohol compared to polystyrene by 2-fold. This improvement in conversion may be attributed to enhanced oxygen solubility. This work demonstrates the importance of oxidant transport and solubility in the nanoreactor core. We are currently investigating methods to improve stability of the nanoreactors to achieve spontaneous phase separation of the product.

526001 Design and Implementation of a Control Strategy Using Risk-Based Probabilistic Design Principles: A Case Study in the Manufacturing an API with a Critical Impurity

Abstract:

In the quality by design (QbD) paradigm proposed by the FDA, the concept of risk analysis and risk management is a fundamental part of the development of the design space and control strategy for critical quality attributes (CQAs) in the manufacture of active pharmaceutical ingredients (APIâ€™s). However, there is not a standardized or well-adopted method for the quantification of risk in the development of processes for the manufacture of API. Herein, we present a case study highlighting the use of probabilistic (Bayesian) modeling techniques as (1) a tool to quantify the relative risk between different processing options, and (2), as a method for developing the control strategy for a specific CQA. We describe the process for manufacture of an API molecule that contains one critical impurity that is controlled as a CQA; the impurity is formed in the final chemical step and is primarily purged in a complex crystallization involving the use of a rotor-stator wet mill. In this case study, using probabilistic models to quantify variability, estimates of expected failure rates for different scenarios were tested, enabling the evaluation of process parameters and different options for control strategies. Lab-scale and pilot plant scale experiments were conducted to generate, test, and validate a series of models, which were used to develop a risk-based probabilistic design space and a control strategy for the CQA. Ultimately, we present the use of probabilistic modeling as a risk-mitigation and optimization tool for the engineer in the pharmaceutical industry.

532026 Proteomics Guided Process Engineering for Anti-HER2 Antibody Production

Abstract:

The high-quality monoclonal antibody (mAb) for cancer treatment is a fast growing market segment of biopharmaceutical industry. Chinese hamster ovary (CHO) cells have been widely used as host cells for mAb production. This study applied a comprehensive proteomics to develop an in-depth understanding of intracellular metabolism regulation and heterologous gene expression, which could benefit the production process optimization. The effect of four basal media and three cell culture operation modes on intracellular protein expression was evaluated in anti-HER2 mAb producing CHO DG44 cell line. A high-titer and high-quality mAb was produced from a fed-batch cell culture process in 2-L bioreactor with titer of >2.5 g/L. Moreover, the possible key host cell regulators of transcription, translation and post translational modification that correlated with important process parameters were identified. Several strategies of rational process engineering were also proposed to guide future cell culture process optimization.

520922 Formation Mechanism and Phase Transformation Behaviors of Polymorphs of Esomeprazole Sodium

Abstract:

Esomeprazole sodium is one kind of proton pump inhibitors. In literature, several polymorphic forms of esomeprazole sodium have been reported. However, no detailed information about the stability, formation mechanism and transformation behaviors of esomeprazole sodium has been reported. In this work, Powder X-ray diffraction (PXRD), thermogravimetric analysis (TGA), differential scanning calorimetry (DSC), Raman spectroscopy, and infrared spectroscopy (IR) were used to identify and characterize different polymorphs of esomeprazole sodium. Single crystal structure of form P was determined. The formation mechanism of form P was investigated and discussed from the molecular level. Presently, form J of esomeprazole sodium is the main commercial product in the market. However, it was found that form J can transform into form P and form L through solvent-mediated transformation and suspension transformation, respectively. Furthermore, selected area electron diffraction (SAED) was used to analysis the structure of different polymorphs of esomeprazole sodium.

525435 Process Modeling of a Continuous Drug Substance Manufacturing Process

Abstract:

Continuous manufacturing has the potential to increase the efficiency, flexibility, agility, and robustness of pharmaceutical processes delivering benefits to both patients and industry. With respect to drug substance manufacturing, there are examples of continuous unit operations such flow synthesis steps for API manufacturing. The current trend is towards increased process integration as opposed to isolated continuous unit operations in an overall batch process and more advanced control strategy approaches including PAT, diversion of non-conforming products etc.

The unit operation connectivity (steady state mode) and effect of disturbances (process dynamic) on process robustness and challenges of process control to produce â€œin-rangeâ€ quality product could be simulated by process modeling. The simulations will be used for case studies and sensitivity analysis to define the sensitivity of the process to particular process parameters, and to investigate the effects of Critical Process Parameters (CPPs) on Critical Quality Attributes (CQAs).

Two steady-state and dynamic process simulation models were developed for an integrated drug substance continuous manufacturing process. The manufacturing process risk analyses were performed and effect of various CPPs on the process robustness and final CQAs were evaluated. With the dynamic model sensitivity analysis and disturbance effects were studied as part of the system dynamic investigation. The system dynamic case studies demonstrate the response of the system to a various disturbances of CPPs (temporal and spatial distributions of disturbances). As a result of this simulation development, process control and manufacturing stability risk analyses were performed and response of the process to different disturbances evaluated under different control strategies.

526637 Green Process Development and Characterization of an Alkylation Reaction to Improve API Impurity Control

Abstract:

Anacetrapib is a CETP inhibitor which was developed to prevent cardiovascular disease, via control of cholesterol levels, by Merck. Extensive efforts have been made to develop an alkylation reaction as the last synthesis step for this drug substance to deliver high conversion, robust impurity control, as well as green chemistry elements such as low Process Mass Intensity (PMI) and reduced waste. Compared to the previously developed NaHMDS-based chemistry, the use of catalytic tetrabutylammonium iodide (TBAI) and potassium carbonate, as a base, not only minimize the epimerization of the final API into the Trans impurity, but also eliminates two process unit operations that were in the previous version of this step. Thorough process characterization was executed with the end-game chemistry to better understand the alkylation reaction, extraction and crystallization procedures for good product quality control. The combination of Design of Experiments (DOE) with statistical analysis and one factor at a time (OFAT) type experiments surrounding multiple process performance-impacting parameters further confirmed the reaction pathway and mechanism. It was found that an undesired hydrolysis reaction and an epimerization of final API were competing against the target pathway. The by-product formation was largely impacted by reactant stoichiometry and TBAI charges, solvent volume and reaction temperature. During the extraction sequence using selected extractants, the hydrophobic nature of API resulted in the formation of a 3rd solution phase (an oily phase), which was found to have no impact on the downstream crystallization process. The risked posed by the oil phase formation in the crystallization process was well controlled by slowing down the batch transfer rate and the final cooldown rate. The optimized reaction, extraction and crystallization process which were performed within the proposed Proven Acceptable Ranges (PAR) for the regulatory filing, showed high conversion, robust impurity control, and improved operability and productivity of the entire process step, as well as decreasing the environmental impact of the step by decreasing the PMI and waste volumes.

529180 Leveraging PAT for Efficient Process Development Workflows

Abstract:

Leveraging PAT for Efficient Process Development Workflows

James C. Marek*, Eric G. Moschetta, Travis B. Dunn,

AbbVie, Inc. Process R&D, 1401 Sheridan Road North Chicago, IL 60064

*speaker: james.marek@abbvie.com

Timelines to develop process understanding and controls for pharmaceutical processes are universally short, necessitating efficient workflows to design and execute data-rich experiments leading to sound conclusions. Leveraging on-line Process Analytical Technology (PAT) is an integral component of efficient workflows. We present a case study that illustrates the power of *in situ* ATR/FTIR to elucidate fundamental kinetics for a complex reaction system, focusing in this talk on the development of the approach. Measuring the relevant chemical species concentrations was essential to modeling the kinetics, and on-line measurements with rapid sampling frequencies collected many data points over the course of the reaction.

The PAT approach was taken after careful consideration of the potential benefits and risks compared to the HPLC method intended for the manufacturing process. Simple calibration models constructed using FTIR measurements of solutions with a single reaction component yielded accuracy sufficient for the purposes of the study. The models were further used to derive species concentrations from FTIR profiles of the reaction run in batch mode. The concentration profiles were readily exported to standard software for kinetic modeling. Because the reaction was relatively fast, and the HPLC method involved a lengthy derivatization step, PAT via on-line ATR/FTIR was clearly the better approach for this investigation. ATR/FTIR was further used to monitor the reaction chemistry in the pilot plant process using the same calibration models developed for the lab study, with results demonstrating excellent scalability of the reaction.

This abstract presentation was sponsored by AbbVie. AbbVie contributed to the design, study conduct, and financial support for the research. AbbVie participated in the interpretation of data, writing, reviewing, and approving the publication. All authors are AbbVie employees.

529073 On-Line FTIR for Accurate Fundamental Kinetic Analysis, Real-Time Process Monitoring, and Process Controls Justification in Pharmaceutical Manufacturing

Abstract:

On-Line FTIR for Accurate Fundamental Kinetic Analysis, Real-Time Process Monitoring, and Process Controls Justification in Pharmaceutical Manufacturing

Eric G. Moschetta,* James C. Marek, and Travis B. Dunn

AbbVie, Inc. Process R&D, 1401 Sheridan Road North Chicago, IL 60064

*corresponding author: eric.moschetta@abbvie.com

Fundamental kinetic analysis of complex reactions is critical to process understanding and development in the pharmaceutical industry. With aggressive timelines for active projects, it is imperative to design and execute data-rich experiments to glean as much kinetic information in as few experiments as possible. However, existing analytical techniques may not be fully developed for complete kinetic analysis, nor may they be amenable to collecting many data points in a single experiment. For example, reaction samples that require labor-intensive workup or processing to analyze quantitatively do not lend themselves well to data-rich experimentation because of the significant burden on the development scientist and the available instrumentation. An alternative approach is to implement on-line process analytical technology (PAT) tools for monitoring the concentrations of the relevant chemical species participating in the reactions. On-line measurements are amenable to collecting many data points over the course of an experiment because the instruments often have rapid sampling frequencies. PAT tools that rely on spectroscopy offer the additional benefit of structural characterization and can distinguish between different chemical species. Careful identification and assignment of characteristic peaks and bands from the on-line signals can provide time-dependent concentration data simultaneously for multiple chemical species. These data can be used not only to build a robust chemometric kinetic model for the process, but for simple identification of logical process endpoints or even to monitor critical quality attributes (CQAs) in real time. As such, implementing on-line PAT tools to improve process understanding at a fundamental level and streamline the workflow in the laboratory is critical to developing robust manufacturing processes in the pharmaceutical industry.

We present a business case in which we implemented an on-line FTIR probe for *in situ* operation in data-rich kinetic experiments. Our business objective was to justify the process controls for the rate of addition of a reagent into a chemical reactor such that the starting material in the reactor was always consumed to the justified level. We hypothesized that a kinetic model describing the decomposition of the starting material would provide excellent justification for the process. However, the kinetic analysis was complicated by two factors: (1) the reagent being dosed to the reactor could further react with the desired product to form an impurity and (2) the current HPLC method required a time- and labor-intensive derivatization process, limiting the amount of data points that could be collected and analyzed in any given experiment. All of our relevant kinetic species contained different functional groups bound to a carbonyl group, so we pursued on-line FTIR as a possible PAT tool to acquire data. We will show how the key experimental observations were more easily elucidated by the on-line IR than the off-line HPLC method and how we used these observations to design and execute the minimum number of data-rich experiments to build a robust kinetic model of the process. Specifically, we will explain how we were able to observe a subtle reaction effect using the on-line FTIR that proved our working hypothesis of the reaction mechanism was wrong. Finally, we show how our chemometric model can be used to monitor pilot plant scale processes in real time based on the fundamental principles developed in the laboratory. This level of understanding allows for predictive power in scalable processes and provides excellent details for process control justification.

The design, study conduct, and financial support for this research was provided by AbbVie. AbbVie participated in the interpretation of data, writing, reviewing, and approving the publication. All authors are AbbVie employees.

529376 From Batch to Continuous Reactive Crystallization: A Case Study on Beta-Lactam Antibiotics

Abstract:

Using data from experiments on batch crystallization to design a continuous system is recognized to be a challenge. Additional complications exist when crystallization is coupled with a chemical reaction and reaction and crystallization kinetics are linked through the generation of a supersaturated environment. Should this coupling of kinetics lead to a Class I behavior (i.e. one in which there is significant supersaturation in the product stream), then the product yield and crystal quality, especially as measured by crystal size distribution, become linked [1].

In the present work, penicillin G acylase (PGA) catalyzes the formation of the antibiotic but also catalyzes the degradation of the product into a slightly soluble by-product. In batch, the PGA-catalyzed reactive crystallization process improved selectivity and productivity by 50% and 25%, respectively [2]. Adapting the batch process for continuous operation can improve productivity further while ensuring better quality consistency, as measured by purity and size distribution. However, PGA must be retained within the crystallizer for the process to be economically feasible, complicating product removal. As a stepping-stone that circumvents this issue, a fed-batch process has been used to tune models accounting for nucleation and growth kinetics, reaction kinetics, and catalyst deactivation kinetics.

Achieving high yields while controlling crystal size distribution requires independent control of the residence times of the solutes (reactants) and solids. Issues encountered include washout, where nucleation or reaction or both are insufficient to maintain a stable slurry; enzyme loss, where catalytic activity diminishes due to either PGA escaping the crystallizer or permanently unfolding; and fines-rich size distributions, where crystal residence time is too short for desired growth. Control over the enzyme reaction rate can be used to reach the desired set point within a feasible space defined by the batch operation. The reaction rate of PGA can be tuned by adjusting temperature, pH, and reactant concentration, however all three of these have been found to impact the antibiotic solubility or crystallization kinetics, further complicating the development of a continuous process. A model of an MSMPR with independent crystal and liquor residence times has been used to optimize the continuous process, and an experimental continuous apparatus has been demonstrated to validate the model while operating under more constrained conditions (imposed by issues of scale and equipment). Compared to the previously published batch process, the *in silico* productivity of the continuous process increases by approximately 2-fold at the sacrifice of approximately 10% of the conversion. The *in silico* average crystal size is small, on the order of tens of microns, comparable to that obtained in batch. The *in vitro* experimental results confirm a decrease in conversion compared to batch; however better purity, consistency, and productivity offset this loss. Experimental crystal sizes were also small, however it may be possible to alleviate this issue with fines destruction.

References

[1] Tavare, N. S. (2013) *Industrial crystallization: process simulation analysis and design*, Springer Science & Business Media.

[2] McDonald, M. A., Bommarius, A. S., and Rousseau, R. W. (2017), *Chem. Eng. Sci. 165*, 81-88.

530097 A New Slurry Reactive Crystallization to Improve Process Robustness and Scalability

Abstract:

Several challenges were faced during the development of development of a drug candidate isolated as a salt. These challenges included high solvent levels entrapped in isolated solids, disproportionation and agglomeration. A root-cause analysis pointed to dimethyl sulfoxide (DMSO), a co-solvent utilized in the crystallization as the main cause for solvent entrapment and agglomeration. DMSO was introduced in the procedure because it enhances drastically the solubility of the drug allowing for a clarifying filtration to be performed at the final step. However, because of its low volatility, DMSO is removed at a slower pace during drying causing partial solubilization of crystal surfaces during drying followed by recrystallization. It is believed that this behavior causes solvent entrapment and agglomeration during drying. In order to address these issues, a study was conducted to identify an alternative co-solvent to DMSO. In this presentation, the development of two types of crystallizations, namely an anti-solvent crystallization and slurry-slurry reactive crystallization to address the issues mentioned above is described. In particular, a new slurry-slurry reactive crystallization, based on an approach recently reported (Derdour et al[â€¡]) was found to be superior in terms of improving process robustness and scalability. That procedure performed successfully when executed at industrial scale.

â€¡ Derdour L., Reckamp J.M. and Pink C., Development of a reactive slurry salt crystallization to improve solid properties and process performance and scalability, Chem Eng Res Des, 121, pp 207-2018, 2017

532745 Crystallization Process Development of Large Size TMP Crystals

Abstract:

Abstract

2,2,6,6-Tetramethyl-4-piperidinol (TMP) crystal is a very important intermediate in organic synthesis. To obtain an improved crystallization yield and desirable crystal-size characteristics, several commonly used organic solvents were investigated to determine the solubility of TMP using a gravimetric method. An ethanol solvent and a cooling crystallization method were chosen to obtain an improved theoretical yield. The cooling crystallization process of TMP in the ethanol solution was further optimized to crystallize larger crystals with a narrow crystal-size distribution using an orthogonal array design. Comprehensive analysis showed that water content in ethanol and seed loading are respectively the main influence factors of the average diameter and the diameter span. Final TMP crystals with an average diameter of 1110 Î¼m and a diameter span of 0.715 were obtained under laboratory conditions using the optimized process.

534105 Polymorph Selection in Continuous Crystallizers in the Presence of Wet Milling

Abstract:

The design of controlled processes that ensure the consistent production of a desired polymorph is an important goal in the context of pharmaceutical manufacturing.[1] The most stable crystalline form can be obtained from a batch crystallization process either through seeding or by allowing enough time for metastable forms to convert to the stable form. On the other hand, producing metastable forms in a batch process relies on delicate process control strategies (temperature, solute concentration and batch time control). As shown in the work by Farmer et al.[4] which polymorph is obtained from a continuous crystallizer at steady state can be controlled by selecting appropriate combinations of residence time, operating temperature and inlet concentration. For cases where the crystallization process is dominated by nucleation and growth, they have shown that the polymorphic outcome in such a continuous process can be accurately predicted once two modified Damköhler numbers are known. In essence, these dimensionless groups represent ratios of the characteristic time for nucleation and growth, the process time (residence time of the crystallizer) and the feed concentration.

In this contribution, we are expanding on this previous works and consider the case where wet milling is used to tune the particle size distribution and potentially alter the polymorphic outcome. We thus present a mathematical model of continuous crystallization with wet milling for polymorphic systems. The model relies on one-dimensional population balance equations including breakage terms. Making the process model dimensionless, we identified a surprisingly small number of additional dimensionless groups that describe the process outcome in terms of polymorphic form when breakage is present. The resulting model is solved using a finite volume scheme (FVS)[5]. Operating regions where the metastable/stable polymorph can reliably be obtained can again be identified. In comparison to the work by Farmer et al.[4], we show that the operating region where the metastable/stable polymorph is obtained can be enlarged by tuning key dimensionless numbers in the breakage kinetics. We combine this type of analysis with an investigation of the productivity of such systems and report regions of attainable particle sizes[4,5]

We show that introduction of a wet suspension mill therefore represents an attractive way to obtain the desired polymorph even if the underlying kinetics of nucleation and growth are disadvantageous. Apart from this finding, we show that introducing breakage increases process productivity and allows to tune particle size more extensively.

[1] Hermanto, M. W.; Chiu, M. S.; Woo, X. Y.; Braatz, R. D. Robust optimal control of polymorphic transformation in batch crystallization. *AIChE J.* 2007,53,2643â€"2650.
[2] Farmer, T. C.; Carpenter, C. L.; Doherty, M. F. Polymorph selection by continuous crystallization. *AIChE J.*2016, 62, 3505â"3514.
[3] Kumar, R.; Kumar, J. Numerical simulation and convergence analysis of a finite volume scheme for solving general breakage population balance equations. *Appl Math Comput.* 2013, 219, 5140-5151.
[4] Vetter, T.; Burcham, C.L.; Doherty, M.F. Regions of attainable particle sizes in continuous and batch crystallization processes, Chem. Eng. Sci., 2014, 106, 167-180.
[5] Vetter, T.; Burcham, C.L; Doherty, M.F. Designing robust crystallization processes in the presence of parameter uncertainty using attainable regions, Ind. Eng. Chem. Res., 2015, 54, 10350-10363.

536947 Polymorph Dynamics of 2-Aminobenzoic Acid System in Continuous Oscillatory Baffled Crystallizer

Abstract:

The polymorph dynamics of 2-Aminobenzoic acid is investigated in a continuous oscillatory baffled crystallizer (COBC). The COBC is a tubular crystallizer containing periodically spaced orifice baffles with oscillatory motion superimposed on the net flow.[1] A Nitech DN15 COBC consisting of eight segments with each segment and elbow jacketed for temperature control is used for the studies. Preliminary experiments show the feasibility of generating the two polymorphs in the COBC via continuous cooling crystallization. However, in order to obtain higher yield combined cooling and antisolvent crystallization (CCAC) is used. In pharmaceutical industry traditionally an antisolvent is added to the solution to increase the yield of the product. However, the presence of the antisolvent can influence the polymorphic outcome.[2] Design of experiments are done to investigate the effects of the parameters – temperature, solvent composition and residence time. The poster will show additional detail studies such as the effect of mixing parameters – piston amplitude and frequency of oscillation on the polymorphic outcome and yield.

[1]Caldeira, R., Ni, X. 2009 Evaluation and Establishment of a Cleaning Protocol for the Production of Vanisal Sodium and Aspirin using a Continuous Oscillatory Baffled Reactor. Org. Process Res. Dev. 13(6), 1080–1087.

[2]Lindenberg, C., Krättli, M., Cornel, J., Mazzotti, M., & Brozio, J. 2008. Design and Optimization of a Combined Cooling/Antisolvent Crystallization Process. Cryst. Growth Des. 9(2), 1124–1136.

535157 In-Situ Optical Imaging and X-Ray Diffraction Techniques to Probe Organic Molecule Thin Film Crystallization

Abstract:

Crystallization of organic materials as thin films is an area of interest in multiple fields, such as in the pharmaceutical industry, organic electronics and for the multiple applications afforded by metal organic frameworks (MOFs). However, many traditional techniques of studying crystal growth and morphology becomes difficult to use for thin film studies, due to the nanometer to micron scale thickness of the films. To overcome this problem, optical and X-ray diffraction techniques can be utilized to probe crystallization of thin films in organic materials. In the crystallization of organic molecules and MOFs, in-situ grazing incidence X-ray diffraction can provide valuable insight in the mechanism of crystal nucleation. In-situ optical studies can then provide complementary information about the crystal growth process in the micron â€" centimeter scale. Together, in-situ optical and X-ray diffraction technique can provide a clear picture of crystallization spanning multiple orders of magnitude in both temporal and spatial dimensions. This talk will focus on the use of in-situ X-ray diffraction and optical imaging to study crystallization from solution of organic molecules and MOFs. The future of in-situ techniques using novel instrumentation will also be discussed.

523533 Imaging Crystallization Using Deep Learning to Quantitatively Track the Polymorphic Transformation of Carbamazepine

Abstract:

Solvent mediated polymorphic transformation (SMPT) is a crucial phenomenon for a specific polymorph crystalsâ€™ production and stability control in pharmaceutical crystallization processes. In this study, a completely new method based on computer vision was developed using deep learning to quantitatively track crystal individuals. For the model compound carbamazepine (CBZ), two polymorphic forms with different morphologies were identified using deep learning image segmentation. The geometry models of two crystal forms were built to calculate the volume of each crystal in the real-time tracking process. The spatial valid vision volume of a high-speed camera probe was calculated and then verified by glass bead. Furthermore, the mass of two polymorphic crystals in suspension were fitted by f_1 and f_2 through comparing real-time tracking and off-line measurement. At last, the real-time quantitative tracking of polymorphic transformation of CBZ was realized through deep learning-based imaging analysis. The transformation process including dissolution of form II, nucleation and growth of form III, crystal size distribution (CSD) etc. was analyzed and identified by off-line measurement. Our work in this study attempts to bridge the gap between the advanced image tracking technology available today and the specific needs of solution crystallization to track, count and measure the crystal individuals. It is a great opportunity to proceed intelligent manufacturing through visualizing traditional crystallization process with the help of artificial intelligence (AI).

Acknowledgement

The authors acknowledge the financial support provided by the Natural Science and Engineering Research Council (NSERC) of Canada, National Natural Science Foundation of China (NNSFC 21576187, NNSFC 21776203, NNSFC 21621004, NNSFC 81361140344), and the China Scholarship Council.

538702 Real Time Constrained Model Predictive Control for the Continuous Manufacturing of Pharmaceuticals

Abstract:

The cost of healthcare in the US is the fastest growing component of the inflation index. One method aimed at reducing these growing costs is in the development of processes for continuous pharmaceutical manufacturing. The continuous production of pharmaceuticals offers reduced costs, simplified and streamlined scale-up, and flexibility compared to batch process alternatives.

This research presents an approach to control system design for the continuous production of pharmaceuticals. First, a dynamic model of the continuous drug manufacturing process is developed that leads to a system of nonlinear, differential algebraic equations. This model is then linearized around typical process operating conditions. The linearized model is subsequently discretized and is used to develop a real time online model predictive control system that ensures the integrity of the pharmaceutical production process and helps satisfy the stringent FDA requirements typically required for safe drug production. Specifically, a constrained infinite time linear quadratic regulator control strategy is employed that can satisfy hard constraints in real time. For example, if there is variability in the quality of raw materials, the controller can ensure, in real time, that FDA requirements on drug purity are met.

The proposed approach is demonstrated with a case study based on a continuous Ibuprofen manufacturing process, as developed for steady-state conditions by Jolliffe & Gerogiorgis (2014), and references therein. In this work, a dynamic model of this process is developed, linearized, and controlled as described above.

534547 Model Predictive Control of a Rotational Molding Process

Abstract:

Rotational Molding, also known as rotomolding or rotational casting, is a batch process used for the processing of plastics through distinct heating and cooling phases [11]. In this process a mold filled with a powdered charge is slowly rotated in a heated oven, resulting in the softened material being dispersed and sticking to the walls of the mold. In order to maintain an even wall thickness, the mold is continuously rotated at all times both during the heating and cooling phase. One of the control challenges is to obtain consistent high-quality products from one batch to another while avoiding unfinished parts with incomplete curing or degradation due to prolonged overheating.

Traditional approach for control of rotational molding processes has been to use an open loop, recipe-based policy. In this approach, a predefined input trajectory is applied to each batch [1]. The approach assumes that the desired product quality can be obtained by repeating historically successful input profiles. Although these approaches are easy to implement and do not require a model for the process, they are incapable of rejecting disturbances that affect the process. This has motivated the use of feedback control strategies.

A classical closed loop control approach that could be used in the rotomolding process is trajectory tracking. In this approach, a predefined set-point trajectory for a measured process variable, such as internal mold air temperature, is tracked. PI controllers are often used for implementing trajectory tracking control. The challenge with trajectory tracking approach is that even with perfect tracking, it may not result in desired quality, as the relationship between the measured/tracked variable and final quality may change significantly with changes in the process conditions across batches.

These challenges are addressed by control strategies that are cognizant of the causal relationship, not only between the manipulated and the (online) measured variables, but also between the manipulated and the final quality variables. A popular model-based control approach, model predictive control (MPC), has increasingly been studied for the control of batch processes [2-5]. The reasons for the popularity of these control schemes are twofold: first, the feedback controller can counter the model uncertainties that are associated with model simplifications and measurement errors; the other being their ability to handle the ever-present input/output constraints. In implementing these formulations, where possible, good first principles models are preferred due to the ability to predict process behavior beyond the data set used and estimate model parameters. However, first principles models are difficult to develop, as it is often impractical to estimate the parameters through experiments, particularly because of the associated cost. Furthermore, these first principles models pose computational challenges for direct use in predictive control formulation, for instance, in case of distributed parameter systems such as batch particulate processes.

These issues have motivated the use of simpler, often linear, models derived from past batch data. A variety of approaches for the development of data-driven models have been proposed. One excellent approach is partial least squares (PLS), which models the process in a projected latent space [6]. These models are essentially time-varying linear models, linearized around mean past trajectories, and therefore require the batches to be the same length, or to recognize an appropriate alignment variable. To account for these limitations, a multi-model approach was proposed in [5]. The multiple models were based on the 'current measurements' of the process instead of the 'time'. These developments were followed by contributions in the area of integration of these data-driven models with the advanced control formulations [4-5]. More recently, a subspace identification based batch control approach was proposed in [7] where a LTI state-space model of the batch process is estimated. Subspace identification methods, in general, are a class of model identification methods that is non-iterative in nature and models a process as a state-space linear time invariant system, identifying the system matrices along with the order of the system. One of the major advantages with this approach is that it facilitates the accommodation of variable duration batches without the need for alignment of batch lengths as demonstrated through different applications in [7-10].

Motivated by these considerations, this work presents a subspace identification based state-space modeling and control approach for rotational molding process to achieve desired quality specifications. Contributions from this developed system represent the progression of this industry towards advanced manufacturing practices, with improved reliability between batches.

[1] D. Bonvin, B. Srinivasan, D. Hunkeler, Control and optimization of batch processes, IEEE Control Systems Magazine 26 (6) (2006) 34-45.

[2] D. Shi, N. H. El-Farra, M. Li, P. Mhaskar, P. D. Christofides, Predictive control of particle size distribution in particulate processes, Chemical Engineering Science 61 (1) (2006) 268-281.

[3] D. Bonvin, Optimal operation of batch reactors- a personal view, Journal of Process Control 8 (5-6) (1998) 355-368.

[4] S. Aumi, P. Mhaskar, Integrating Data-Based Modeling and Nonlinear Control Tools For Batch Process Control, AIChE Journal 58 (2012) 2105-2119.

[5] S. Aumi, B. Corbett, P. Mhaskar, and T. Clarke-Pringle. Data-based modeling and control of nylon-6, 6 batch polymerization. Control Systems Technology, IEEE Transactions on, 21(1):94-106, Jan 2013.

[6] J. Flores-Cerrillo, J. F. MacGregor, Control of particle size distributions in emulsion semibatch polymerization using mid-course correction policies, Industrial and Engineering Chemistry Research 41 (7) (2002) 1805-1814.

[7] B. Corbett, P. Mhaskar, Subspace identification for data-driven modelling and quality control of batch processes, AIChE Journal 62 (2016) 1581-1601.

[8] Abhinav Garg, P. Mhaskar (2017), Subspace Identification Based Modeling and Control of Batch Particulate Processes, Industrial and Engineering Chemistry Research, 56 (26) 7491-7502.

[9] Abhinav Garg, Brandon Corbett, Prashant Mhaskar, Gangshi Hu, Jesus Flores-Cerrillo (2017), High Fidelity Model Development and Subspace Identification of a Hydrogen Plant Startup Dynamics, Computers and Chemical Engineering, 106 183-190.

[10] Rashid, M. M., Mhaskar, P., & Swartz, C. L. (2017). Handling multiâ€rate and missing data in variable duration economic model predictive control of batch processes, AIChE Journal, 63(7), 2705-2718.

[11] Felipe P.C. Gomes, Abhinav Garg, P. Mhaskar and Michael R. Thompson, Quality monitoring of rotational molded parts using a nondestructive technique, ANTEC (2018), Accepted.

534349 A Multiparametric Programming Based Approach to Integrate Design, Scheduling, and Control of a Batch Process

Abstract:

Integrating short term (control), middle term (schedule), and long term (design) decisions has been shown to improve the operability and profitability of batch processes [1]. The interactions between these distinct time-scale decisions allow for opportunities to (i) replace fixed recipe-based schedules with detailed model-based schedules, (ii) build design dependent schedule aware controllers, and (iii) reduce capital costs by circumventing overdesigned processing units. While the efficiency of simultaneous decision making in process systems is greatly acknowledged, the methodology/procedure to achieve an effective integration is still an open question due to the order of magnitude differences in the time scales of the constituent problems, as well as their respective objectives. The most recent advances within the last decade [2-7] depict a promising basis for further improvement.

In this work, we present a unified framework to simultaneously determine the optimal codependent short term and middle term operating strategies of a batch process for a range of market conditions, while minimizing the capital costs. A single high fidelity model is utilized to develop (i) multiparametric rolling horizon optimization (mpRHO) strategies as a function of process states and time variant market conditions, and (ii) multiparametric Model Predictive Control (mpMPC) for effective set-point tracking, following a similar methodology described in Burnak *et al* [8]. The design of the process is accounted for as an additional unknown bounded parameter in the offline operational strategies. The offline nature of mpRHO and mpMPC strategies allows for integration in the design optimization of the process, i.e. minimizing the capital costs. The framework is showcased on (i) a single non-isothermal batch reactor, and (ii) a batch process comprising a preprocessing unit, a reactor, and a post-processing unit.

1. Pistikopoulos, E. N., Diangelakis, N. A., Towards the integration of process design, control and scheduling: Are we getting closer?, Computers and Chemical Engineering (2015), 91, pp. 85-92.
2. Zhuge, J., Ierapetritou, M. G., Integration of scheduling and control for batch processes using multi-parametric model predictive control, AIChE J. (2014), 60 (9), pp. 3169-3183.
3. Baldea, M., Harjunkoski, I., Integrated production scheduling and process control: A systematic review, Computers and Chemical Engineering (2014), 71, pp. 377-390.
4. Dias, L. S., Ierapetritou, M. G., Integration of scheduling and control under uncertainties: Review and challenges, Chemical Engineering Research and Design (2016), 116, pp. 98-113.
5. Patil, B.P., Maia, E., Ricardez-Sandoval, L.A., Integration of scheduling, design, and control of multiproduct chemical processes under uncertainty, AIChE Journal (2015), 61, pp. 2456-2470.
6. Dias, L. S., Pattison, R. C., Tsay, C., Baldea, M., Ierapetritou, M. G., A simulation-based optimization framework for integrating scheduling and model predictive control, and its application to air separation units, Computers and Chemical Engineering (2018), pp. 139-151.
7. Rossi, F., Casas-Orozco, D., Reklaitis, G., Manenti, F., Buzzi-Ferraris, G., A computational framework for integrating campaign scheduling, dynamic optimization and optimal control in multi-unit batch processes, Computers and Chemical Engineering (2017), 107, pp. 184-220.
8. Burnak, B., Katz, J., Diangelakis, N. A., Pistikopoulos, E. N. Simultaneous process Scheduling and control: A multiparametric programming based approach, Industrial & Engineering Chemistry Research (2018), 57 (11), 3963-3976.

524393 Multi-Objective Dynamic Optimisation of a Fed-Batch Nosiheptide Reactor

Abstract:

Thiopeptide antibiotics are highly potent and structurally complex; this growing class of secondary metabolites are formed by ribosomal peptide biosynthesis. Nosiheptide, a sulfur-containing peptide antibiotic obtained through fermentation, can be regarded as the structural prototype of the class e thiopeptide antibiotics. It exerts exceptional antibiotic activity in vitro and in a mouse model against critical Gram-positive pathogens such as MRSA, VRE, or *Clostridium difficile*. Shown non-toxic at high dosages it has previously been used for applications in farm animals (Benazet at al., 1980). Recently the first total synthesis of nosiheptide was reported, utilizing double macrocyclization of a fully functionalized linear precursor (Wojtas et al., 2016). Given low indusial yields, strong motivation exists to dynamically optimise the process for improved product yield while reducing production time and cost. Similar multi-objective optimisation problems are common in the chemical industry (Rodman, Fraga and Gerogiorgis, 2017), which can be solved with both stochastic and deterministic techniques.

The nosiheptide fermentation process is an extremely complex biochemical reaction, with the complete reaction mechanism not strictly understood. A simplified mechanistic fermentation process model has been proposed, considering only the most essential interactions in the biochemical network (Wojtas et al., 2016). The model considers the temperature and pH dependence of biomass growth, in addition to substrate consumption and product production, with the 19 model parameters determined from the model authors experimental campaigns.

To circumvent mass transfer inhibition at excessive substrate concentrations in the broth the nosiheptide reactor is operated in a fed-batch mode, where the reactor is only partially filled initially, with substrate supplemented over time. Herein the multi-objective dynamic optimisation problem can be explicitly formulated: our two minimisation criteria are batch time and inverse yield with four decision (control) profiles to compute towards optimising the system: both the dynamic feed flow rate and the dissolved oxygen concentration must be determined in addition to the reactor temperature and pH profile subject to operability constraints.

This contribution employs (epsilon constrained ***or*** variable weighted sum) handling of the bi-objectives problem, allowing the Pareto front trade-off between the competing objectives to be visualised. A direct method for dynamic optimisation (simultaneous strategy) has been performed in each case to compute the four optimal control trajectories. Orthogonal polynomials on finite elements are used to approximate the control and state trajectories allowing the continuous problem to be converted to NLP form (Biegler, Cervantes, and Wächter, 2007). The resultant large scale NLP problem is been solved for each instance using IPOT (Wächter and Biegler, 2006), and global optimality is ensured with a multi-start search via Latin Hypercube Sampling of the input space for initialisation.

The algorithm employed demonstrates that appropriate reactor operation and control can significantly improve product yield. Additionally, explicit visualization of the objective trade-offs provides valuable insight to decision makers towards critical economic process decisions.

REFERENCES

Benazet, F., Cartier, M., Florent, J., et al., 1980. Nosiheptide, a sulfurcontaining peptide antibiotic isolated from Streptomyces actuosus 40037. *Cell. Mol. Life Sci.* **36**(4): 414â"416.

Biegler, L. T., Cervantes, A. M. and WÃ¤chter, A., 2002. Advances in simultaneous strategies for dynamic process optimization. *Chemical Engineering Science* **57**: 575-593.

Biegler, L.T., 2007. An overview of simultaneous strategies for dynamic optimization. *Chemical Engineering and Processing: Process Intensification* **46**(11): 1043-1053.

Niu, D., Jia, M., Wang, F. and He, D., 2013. Optimization of nosiheptide fed-batch fermentation process based on hybrid model. *Industrial & Engineering Chemistry Research* **52**(9): 3373-3380.

Rodman, A.D. and Gerogiorgis, D.I, 2017. Dynamic optimization of beer fermentation: sensitivity analysis of attainable performance vs. product flavor constraints. *Computers and Chemical Engineering* **106**(C): 582-595.

Rodman, A.D., Fraga, E.S. and Gerogiorgis, D.I., 2018. On the application of a nature-inspired stochastic evolutionary algorithm to constrained multi-objective beer fermentation optimisation. *Computers and Chemical Engineering* **108**: 448-459.

Wojtas, K.P., Riedrich, M., Lu, J.Y., Winter, P., Winkler, T., Walter, S. and Arndt, H.D., 2016. Total Synthesis of Nosiheptide. *Angewandte Chemie International Edition*, **55**(33): 9772-9776.

534565 Big Data and the Chemical Industry: From Sensor to Value

Abstract:

With the Internet of Things, smart and wireless sensors, advanced control systems, wireless communications and smart devices, the amount of data collected and stored has grown exponentially. These bring big opportunities to the chemical industry where Big Data can be used to enhance manufacturing & operational excellence, forecast pricing, improve formulations, and improve marketing and support innovation.

We present the challenges and opportunities of Big Data in SABIC and a case study illustrating the use of BD technology to move from data, to knowledge, to value in support of Innovation.

536696 Big Data and Iot: A Demonstration Testbed of Multi-Stage Centrifugal Pumping System

Abstract:

Big data and IoT: a demonstration testbed of multi-stage centrifugal pumping system

Devarshi Shah[1], Q. Peter He[1], and Jin Wang[1]

[1]Department of Chemical Engineering, Auburn University, Auburn, AL, 36849, USA

Abstract:

A new industrial revolution is happening to make current manufacturing systems smarter, safer, and more efficient. With the emergence of the Industrial Internet of Things (IoT) and ever advancing computing power and expansion of wireless networking technologies, a new generation of networked, information-based technologies, data analytics, and predictive modeling are providing unprecedented embedded computing capabilities as well as access to previously unimagined potential uses of data and information. This revolution is known as smart manufacturing [1] or Industry 4.0.

Although many general frameworks have been proposed for IoT enabled systems for industrial application [1]-[3]. There is limited literature on demonstrations or testbeds of such systems. In addition, there is a lack of systematic study on the characteristics of IoT sensors and data analytics challenges associated with IoT sensor data. This study is an attempt to help fill this gap by exploring the characteristics of IoT vibration sensors and show how IoT sensors and big data analytics can be used to develop real time monitoring frameworks.

In this work, we will introduce the design of an IoT testbed using multi-stage centrifugal pumping system equipped with non-invasive IoT vibration sensors as shown in Fig. 1. We will discuss the system architecture and protocols designed for data collection, transmission, and storage as shown in Fig. 2. We will also discuss the characteristics of IoT sensors and data analytics challenges associated with data collected from these sensors. The main focus of this work is the development of data-driven predictive models based on the vibration signals to infer flowrate inside the pipe and rpm of the pump motor, and compare with experimental measurements. Altogether, this study serves as a demonstration of how IoT sensors and big data analytics can be integrated and utilized for real-time process monitoring.

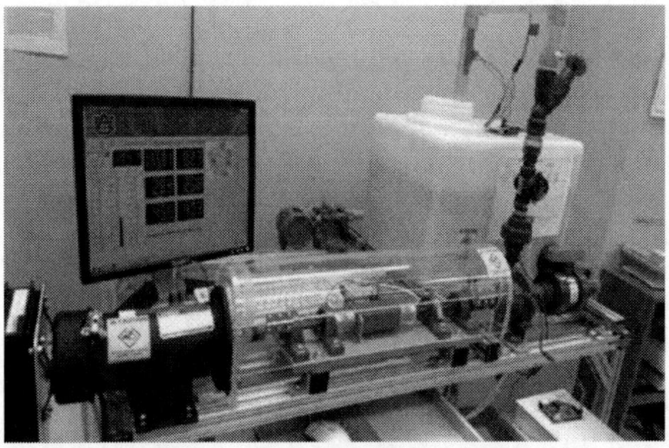

Fig. 1 Actual setup of the testbed with five IoT vibration sensors attached to the pumping system at five different locations

References:

[1] J. Davis, T. Edgar, J. Porter, J. Bernaden, and M. Sarli, "Smart manufacturing, manufacturing intelligence and demand-dynamic performance," *Comput. Chem. Eng.*, vol. 47, pp. 145–156, Dec. 2012.

[2] G. Fortino and P. Trunfio, Eds., *Internet of Things Based on Smart Objects*. Cham: Springer International Publishing, 2014.

[3] H. S. Kang, J. Y. Lee, S. Choi, H. Kim, J. H. Park, J. Y. Son, B. H. Kim, and S. Do Noh, "Smart manufacturing: Past research, present findings, and future directions," *Int. J. Precis. Eng. Manuf. Technol.*, vol. 3, no. 1, pp. 111–128, Jan. 2016.

536078 Pressure Swing Adsorption Cycle Synthesis Utilizing Artificial Neural Networks As Surrogate Models

Abstract:

With fossil fuels expected to be a significant portion of the worldâ€™s energy mix for the near future, it is important to minimize CO_2 emissions from current power plants through carbon capture and sequestration (CCS).[1] In post-combustion CCS, CO_2 is separated from the power plant flue gas emissions, containing mainly of N_2 and CO_2. Of the technologies currently available for CCS, pressure swing adsorption (PSA) is one of the more promising due to low energy requirements and short cycle times compared to other adsorption based technologies. However, one challenge that exists in using PSA for CCS is designing the cycle to match newly developed materials for this application. Although the steps in all PSA cycles can be classified into six different possibilities (pressurization, feed, depressurization, light reflux, heavy reflux and pressure equilibration), the arrangement of the steps and interactions between steps lead to hundreds of potential different combinations. The objective of this work is to develop a new approach that is capable of synthesizing PSA cycles to capture the CO_2 from flue gas at the required purities and recoveries while minimizing energy requirements and maximizing adsorbent productivities.

In this work, we present a new framework for synthesizing the PSA cycle with the lowest predicted CO_2 capture costs. In this framework, we train artificial neural networks (ANNs) using Bayesian regularization methods as surrogate models for the various PSA steps. The ANNs are trained on simulation data collected from our PSA model consisting of a system of partial differential algebraic equations incorporating mass and energy balances, pressure drop across the column, competitive multi-site Langmuir isotherms and the linear driving force model.[2] With the ANN surrogate models, we propose a mixed integer nonlinear programming (MINLP) model to determine the ideal ordering and duration of steps in order to minimize the energy requirements and maximize the adsorbent productivity. We evaluate this model with several adsorbents, including Ni-MOF-74, UTSA-16 and zeolite 13X to compare the adsorbents under optimized cycle conditions.

1. MIT. *The Future of Coal*; 2007. http://web.mit.edu/coal/.
2. Leperi KT, Snurr RQ, You F. Optimization of Two-Stage Pressure/Vacuum Swing Adsorption with Variable Dehydration Level for Postcombustion Carbon Capture. *Ind. Eng. Chem. Res.* 2016;55:3338-3350.

537187 Continuous Control of a Polymerization System with Deep Reinforcement Learning

Abstract:

Improving and deriving noble control laws for polymerization reaction has been one of the uppermost tasks in the chemical industry. There have been extensive studies in control industries from proportional controls to various forms of model predictive controls, which have been benefiting in current chemical processes and plants[1]. Recent breakthroughs in deep learning area have started to inspire the development of Artificial Intelligence(AI)-based controllers in chemical reaction control area thanks to its known success in wide-range applications from gaming to robotics control[2]. These applications utilize deep reinforcement learning (Deep RL), of which the ultimate goal is to enable computers to make human-like policies based on AI's exploration of the environment[3]. However, some challenges are still faced for applying Deep RL in controlling chemical reactions, since the inputs of chemical reactors are usually high-dimensional, and the system dynamics are often sensitive and has considerable time-delay effect[2].

Previous studies have been done in controlling a free-radical polymerization process by following real-time measurement of weight-average molar mass[1], where the specific molar mass distribution can be achieved by following a trajectory of weight-average molar mass with respect to time[1]. In this work, we developed a deep learning based controller for a free radical poly-acrylamide polymerization system using Deep Deterministic Policy Gradient (DDPG). The DDPG utilizes actor-critic structure that is able to predict actions of infinite dimensions[4]. The controller calculates the control action a_t, which consists of the monomer flow rate F_m and initiator flow rate F_i to adjust at each time step t, and the system response of the action is recorded as a state s_t. The network is trained to maximize the cumulative reward r that accounts for the distance between current output and target output for each iteration[5]. The network is trained on an established kinetic model of the polymerization reaction, and the controller gradually learns the policy through exploration of the system. Then convergence is achieved when the average cumulative reward reaches a desired threshold. In our experiment, the controller successfully has learned the control policy to follow the target trajectory of the weight-average molar mass.

Overall the smart controller has shown robust control over a range of operating conditions, which indicates the deep reinforcement learning based approach's capability in controlling a nonlinear dynamic semi-batch system.

Reference

[1] N. Ghadipasha, W. Zhu, J. A. Romagnoli, T. Mcafee, T. Zekoski, and W. F. Reed, "Online Optimal

Feedback Control of Polymerization Reactors : Application to Polymerization of Acrylamide – Water – Potassium Persulfate (KPS) System," 2017.

[2] S. P. K. Spielberg, R. B. Gopaluni, and P. D. Loewen, "Deep Reinforcement Learning Approaches for Process Control", 2017

[3] V. Mnih, D. Silver, and M. Riedmiller, "Playing Atari with Deep Reinforcement Learning," pp. 1–9.

[4] T. P. Lillicrap *et al.*, "Continuous learning control with deep reinforcement learning," , 2016.

[5] D. Silver, G. Lever, D. Technologies, G. U. Y. Lever, and U. C. L. Ac, "Deterministic Policy Gradient Algorithms."

524390 Process Modelling, Simulation and Optimisation for Continuous Biopharmaceutical Manufacturing

Abstract:

Batch processing is the traditional mode of manufacturing for a wide range of products by the pharmaceutical industry. While batch methods have advantages such as flexible equipment usage and the ability for specific batch recall, issues with product quality variations and process reliability necessitate technological innovation to maintain profitability and sustainability. Continuous pharmaceutical manufacturing (CPM) is a promising alternative to the batch paradigm [1], offering remarkable operating and economic benefits [2], attracting strong interest from academia, industry and regulatory bodies [3]. Continuous biopharmaceutical manufacturing also shows promise for technoeconomic benefits [4]; the elucidation of end-to-end continuous biopharmaceutical manufacturing campaigns, including essential continuous purification and separation processes, is essential for successful implementation.

Process systems engineering (PSE) methodologies are valid alternatives to laborious experimental campaigns at the early stages of process development, allowing screening for promising candidate process configurations prior to further development and scale-up [5], as well as in retrofitting [6]. For sterile filling of biopharmaceuticals, a systematic decision-making procedure has been developed to support the selection of equipment technology, in order to choose between conventional multi-use or novel single-use options [7]. Process modelling and nonlinear optimisation studies have also been used to establish cost optimal operating configurations for the CPM of pharmaceutical [8] and biopharmaceutical products [9]. However, plantwide modelling, simulation and technoeconomic optimisation of a continuous biopharmaceutical manufacturing plant has yet to be implemented to establish such cost optimal operating configurations.

This paper presents the nonlinear optimisation for total cost minimisation of the upstream continuous manufacturing of a biopharmaceutical product. Kinetic parameter estimation for biopharmaceutical synthesis modelling has been performed on the basis of available literature data for a batch culture of hybridoma cell line [10] and for a continuous culture of budding yeasts [11]. Plantwide models for both upstream synthesis and downstream separation process models are then discussed, and the formulation of the objective function (plant cost) for total cost minimisation is presented, along with the illustration of minima for various process configurations, followed by a critical discussion of the implemented methodology. This work illustrates potential advantages attainable via end-to-end continuous biopharmaceutical manufacturing, and the importance of process modelling and optimisation studies at the early stages of design.

REFERENCES

[1] B. Gutmann, D. Cantillo, and C. O. Kappe, "Continuous-Flow Technology: A Tool for the Safe Manufacturing of Active Pharmaceutical Ingredients," *Angew. Chemie-International Ed.* **54**(23): 6688–6728 (2015).

[2] D. Dallinger and C. O. Kappe, "Why flow means green – Evaluating the merits of continuous processing in the context of sustainability," *Curr. Opin. Green Sustain. Chem.***7**: 6–12 (2017).

[3] S. L. Lee, T. F. O'Connor, X. Yang, C. N. Cruz, S. Chatterjee, R. D. Madurawe, C. M. V. Moore, L. X. Yu, and J. Woodcock, "Modernizing Pharmaceutical Manufacturing: from Batch to Continuous Production," *J. Pharm. Innov.* **10**(1): 191–199 (2015).

[4] M. S. Croughan, K. B. Konstantinov, and C. Cooney, "The future of industrial bioprocessing: Batch or continuous?," *Biotechnol. Bioeng.* **112**(4): 648–651, (2015).

[5] S. K. Teoh, C. Rathi, and P. Sharratt, "Practical Assessment Methodology for Converting Fine Chemicals Processes from Batch to Continuous," *Org. Process Res. Dev.* **20**(2) 414–431 (2015).

[6] A. Segawa, S. Yoshikawa, T. Toyama, H. Nakanishi, E. Kikuchi-Uehara, M. Hirao, and H. Sugiyama, "Method for reducing environmental, health, and safety risks in active pharmaceutical ingredient manufacturing based on multiobjective evaluation," *Process Saf. Environ. Prot.* **104**: 304–313 (2016).

[7] H. Shirahata, M. Hirao, and H. Sugiyama, "Decision-Support Method for the Choice Between Single-Use and Multi-Use Technologies in Sterile Drug Product Manufacturing," *J. Pharm. Innov.* **12**(1): 1–13 (2017).

[8] H.G. Jolliffe, S. Diab, and D. I. Gerogiorgis, "Nonlinear optimisation via explicit NRTL model solubility prediction for antisolvent mixture selection in artemisinin crystallisation," *Org. Process Res. Dev.* **22**(1): 40–53 (2018).

[9] P. Kroll, A. Hofer, S. Ulonska, J. Kager, and C. Herwig, "Model-Based Methods in the Biopharmaceutical Process Lifecycle," *Pharm. Res.* **34**(12): 2596–2613 (2017).

[10] C. Kontoravdi, E. N. Pistikopoulos, and A. Mantalaris, "Systematic development of predictive mathematical models for animal cell cultures," *Comput. Chem. Eng.* **34**(8): 1192–1198 (2010).

[11] H. N. Chang, N. J. Kim, J. Kang, C. M. Jeong, J. D. R. Choi, Q. Fei, B. J. Kim, S. Kwon, S. Y. Lee, and J. Kim, "Multi-stage high cell continuous fermentation for high productivity and titer," *Bioprocess Biosyst. Eng.* **34**(4): 419–431 (2011).

523979 Process Control Strategies for Pharmaceutical Drug Product Continuous Manufacturing – Where Are We Now and Where Do We Go from Here?

Abstract:

Drug product continuous manufacturing is approved for multiple products. It is no longer new but it is early in the technology lifecycle and is rapidly evolving. The talk will be organized around the concept of process control as a triad of "measure, model, and control" to assure the highest product quality at the lowest cost. It will review the different approaches being taken to process control currently and discuss opportunities for future innovations in measurement, real-time modeling, and control.

532537 Data Accuracy for Smart Manufacturing in Continuous Pharmaceutical Systems

Abstract:

The principles of Quality by Design (QbD) and the innovative applications of Process Analytical Technology (PAT) tools have enabled the adoption of continuous processing paradigms to advance pharmaceutical manufacturing. As the community progresses towards strategies for real-time release testing and implementing advanced process control strategies, it is important to ensure accuracy in process data. Process and material quality measurements using PAT tools are highly valuable, however, present multiple challenges such as extensive calibration, data handling, sensor fouling, measurement bias, measurement drifts etc. The utility of such tools in real time hence can become limited unless properly managed. However, by leveraging the extensive real time generated by robust equipment sensors and field devices along with predictive capabilities offered by process models, the application of data reconciliation and gross error detection can enable enhanced real-time process data accuracy and validation [1]. Improved accuracy in raw measurements enhances plant-enterprise integration with increased reliability.

Data reconciliation (DR) and gross error detection (GED) are systems engineering tools for further improving the accuracy and consistency of process data through systematic data rectification and checking [2]. DR can be posed as a constrained optimization problem, which has at its objective providing the best estimate of process variables, which satisfy the process model, including material and energy balances. Under DR random errors and gross errors associated with measurements are rectified through a model-based approach providing the sensor network has the necessary redundancy. Reliable state estimates of process variables not only enhances process monitoring capabilities but also improves the performance of control systems.

DR is a powerful tool for continuous processes whose application in industries such as oil and gas is well known. It is a powerful tool for leveraging large datasets for improving monitoring consistency in the pharmaceutical industry. In this paper, we demonstrate improved accuracy in real-time monitoring of relevant process variables in a continuous dry granulation tableting line. The dry granulation process exhibits nonlinear response and fast dynamics and hence poses interesting challenges for monitoring and control. In this paper, an earlier simulation study highlighting the application of DR for a partial dry granulation line [3] is extended to investigate experimental performance using the entire tableting line. The DR is conducted using a steady-state model in order to achieve fast computational times. The DR framework is experimentally demonstrated using real time measurements of the relevant CQA's on the continuous tableting pilot plant at Purdue University.

References

[1] M. J. Bagajewicz, Smart Process Plants: Software and Hardware Solutions for Accurate Data and Profitable Operations, New York: McGraw-Hill, 2010

[2] S. Narasimhan and C. Jordache, â€œThe Importance of Data Reconciliation and Gross Error

Detection,â€ in Data Reconciliation and Gross Error Detection, Elsevier, 1999, pp. 1â€“31.

[3] S. Ganesh, M. Moreno, J. Liu, M. Gonzalez, Z. Nagy, G. Reklaitis. Sensor Network for Continuous Tablet Manufacturing, Proceedings of 13th International Symposium on Process Systems Engineering, 2018 (accepted)

531857 Impurity Control in the Continuous Reactive Crystallization of Beta-Lactam Antibiotics

Abstract:

Beta-lactam antibiotics such as cephalexin and ampicillin can be synthesized and crystallized simultaneously with the use of penicillin G acylase (PGA) as a catalyst. However, PGA also catalyzes the degradation of the antibiotic to form a slightly soluble by-productâ€"phenylglycine in the case of cephalexin or ampicillinâ€"that can contaminate the solid product [1]. It is important that by-product concentration remain below the solubility limit so that pure antibiotic can be filtered immediately without the need for recrystallization or other further purification. In a continuous process, online detection of phenylglycine crystals is necessary to ensure product quality. Focused beam reflectance measurement (FBRM) has been used to observe the nucleation of the byproduct in real time. However, it is desirable to take preemptive action to avoid nucleation of the by-product at all. The combination of several process analytical technologies such as ATR-FTIR and inline polarimetry enable the detection of phenylglycine before the solubility limit is reached. Purity can then be enforced by changing the crystallizer conditions to increase enzyme selectivity or decrease enzyme activity, both at a cost to productivity and conversion, but without the need to stop the continuous process due to solid phase impurity. A model of the reactive crystallization system is used to inform controller actions. Experiments conducted in a fed-batch crystallizer, where phenylglycine accumulation is easier to control, are used to evaluate phenylglycine crystal detection by FBRM as well as model accuracy. A mixed-suspension mixed-product removal (MSMPR) crystallizer is used to evaluate different control actions, such as decreasing pH (which simultaneously decreases antibiotic solubility and increases PGA selectivity), changing temperature (increasing temperature decreases solubility of phenylglycine but increases PGA activity while decreasing temperature has the opposite effect), and changing feed reactant concentrations (changing the reactant ratio affects both selectivity and activity).

References:

[1] McDonald, M.A., Bommarius, A.S., and Rousseau, R.W. (2017) *Chem. Eng. Sci. 165*, 81-88.

532353 Continuous Processing of Complex Dosage Forms: Liposomes and Polymeric Micelles

Abstract:

PURPOSE

Complex dosage forms such as liposomes and polymeric micelles have tremendous potential to revolutionize the pharmaceutical industry and provide significant health benefits. One challenge is to process large quantities of these nanoparticles while also maintaining physicochemical properties to ensure a high quality formulation. One approach to overcome such challenge was to turn to continuous processing. Continuous processing can achieve improved process control (*via* process analytical technology), while avoiding inherent issues associated with batch processing to ultimately improve product quality and reduce manufacturing cost. Accordingly, we developed a continuous process for liposomes and polymeric micelles. In this work, we outline important steps in the particle formation process, in-line/ at-line measurements, in-line concentrating and drug-loading strategies.

METHODS

Risk Assessment

Ishikawa diagrams were formed to outline process parameters and material attributes that had the likelihood to effect quality attributes. Some common critical quality attributes (CQAs) for nanoparticle formulations include particle size distribution (z-average particle size and polydispersity index), zeta-potential, nanoparticle concentration and encapsulation efficiency. A custom-built system was designed to incorporate process control strategies to ensure CQAs were within user specifications.

Nanoparticle Formation Process

The nanoparticles were formed using an ethanol injection method with a custom-built continuous processing system that is controlled using a program written using National Instruments LabVIEW. Briefly, the continuous processing system consists of multiple tanks that contain lipid dissolved in ethanol and another set of tanks for the aqueous components (*e.g.* DI water or buffer salts). The system further consists of an injection port such that a turbulent jet is formed under controlled flow conditions with downstream conditioning modules and concentrating modules.

In-line/ at-line measurements include: (1) flow rates; (2) temperatures; (3) particle sizing *via* dynamic light scattering; (4) particle concentration *via* an NIR turbidity meter; (5) pH and (6) conductivity.

Active-loading Doxorubicin-HCl into Liposomes
Hydrogenated soy phosphatidyl-choline (HSPC), cholesterol and DSPE-mPEG2000 were prepared at a molar ratio of 5.6:3.8:0.52 and were dissolved in ethanol. The lipid solution was subsequently injected into the aqueous phase (250 mM Ammonium Sulfate) *via* the co-flow injection port. Liposomes were then concentrated downstream and doxorubicin-HCl was injected into the pre-formed liposomes under controlled conditions. Offline UV/HPLC measurements were performed to confirm loading percentage.

RESULTS

HSPC:Chol:DSPE-mPEG2000 liposomes were successfully formed under different processing conditions. The z-average particle size could be controlled from approximately 40.0 to >100 d.nm with a PDI of less than 0.10 under most conditions. For the doxorubicin loading studies, high encapsulation efficiency was achieved using the liposomes prepared on our continuous processing system.

CONCLUSION

Continuous processing of liposomes that incorporates a turbulent jet in co-flow enables a fine-control of nanoparticles. On-line/ at-line measurements techniques are important to characterize and provide feedback during the continuous process and can easily be implemented into a process control strategy. Lastly, we have provided groundwork on continuous processing of liposomes that can be used to make doxorubicin-loaded liposomes with similar quality attributes of existing drug products.

ACKNOWLEDGEMENTS

This work was supported by the FDA (HHSF223201610121C and 1U01FD005773-01).

DISCLAIMER

This article reflects the views of the authors and should not be construed to represent FDAâ€™s views or policies.

534546 Active Process Control in the Quality-By-Design (QbD) Implementation of Pharmaceutical Continuous Tablet Manufacturing

Abstract:

Continuous manufacturing has been routinely implemented and systematically improved in the chemical, petroleum and refinery industries. In the last decade, the stringently regulated pharmaceutical industry has undergone a shift in paradigm from the conventional batch manufacturing to continuous manufacturing. [1] Under this trend, real-time product quality assurance in pharmaceutical continuous manufacturing (PCM) has been widely investigated under the Quality-by-Design (QbD) guidelines recommended by the US Food and Drug Administration. Active process control was highlighted recently [2] as a key component to enable QbD implementation. Specifically, it was proposed that the active control of critical quality attributes (CQA), such as drug loading in oral solid dosage, can address variations in critical material properties (CMA), such as bulk density of active pharmaceutical ingredients (APIs) by manipulating critical process parameters (CPP), such as tablet press dosing position. Despite the feasibility of implementing active process control, which has been demonstrated for a number of continuous manufacturing facilities, including continuous direct compaction, there remains hesitation in widespread adoption of active process control strategies due to existing investments in established and mature batch technologies. [3] In this work we investigate the development of an efficient active process control design for a rotary tablet press.

A rotary tablet press has a number of punch stations (e.g., 16) mounted on a round turret. Each punch station undergoes the following steps: die filling and metering, pre-compaction, main-compaction, tablet ejection, and tablet take-off from lower punch. Conventional batch manufacturing employs pre-batch testing of in-process CQAs, such as API composition and mixing uniformity of the blend, and post-batch testing of tablet CQAs, such as weight and tensile strength. This operating procedure minimizes the importance of active process control and thus requires CPPs (such as main compression force, dosing position, turret speed) to be kept within a predefined design space during the entire run. By contrast, continuous manufacturing relies on active process control, under QbD guidance, to ensure consistent quality production in real time and with less trial and error testing. Thus, active process control is a key component to enabling widespread adoption of pharmaceutical continuous manufacturing. However, an efficient active process control system design for rotary tablet press has not been fully investigated and its full benefits have not been elaborated. Specifically, the advantages of active process control algorithms and structures for rotary tablet press over established and mature batch manufacturing strategies in terms of product quality variances or risks merit investigation. It is worth noting that the robust design of traditional manufacturing equipment (such as a rotary tablet press, a roller compactor, etc.) has resulted in minimum variation of CPPs and CQAs during operation and thus allowed batch pharmaceutical manufacturing to assess quality using post-batch statistical quality control (SQC) methods. An active process control system, by contrast, can use possibly noisy and biased CQA measurements to effectively supervise the control of CPPs and thus minimize the need for batch-end SQC. An important aspect to investigate is the extent to which this integration imposes additional dynamics on the process, and how this could potentially amplify variations in CPPs and thus in CQAs.

In this study, a hierarchical three-level active process control, designed according to the ANSI/ISA 95 standard, was implemented on a Natoli BLP-16 rotary tablet press integrated within a continuous direct compaction pilot plant. [4] A Level 0 control consisted of the CPP controls built-into the programmable logic control (PLC) panel provided by the equipment vendor for CPPs (turret speed, dosing position, etc.). A Level 1 control was employed at the OPC server level using cascaded PID close-loop control of CQAs (drug loading, tablet weight, and tensile strength) by means of tablet weight and API composition measurements, obtained with load cell and NIR spectroscopy in-house designed sensors, respectively. A Level 2 controller was designed, using data reconciliation (DR) and model predictive control (MPC) techniques, with the objective of reconciling and supervising the control of CQAs and CPPs at the lower levels. In addition, an at-line Sotax AT4 tester was employed to collect a string of 16 tablets every five minutes with the purpose of determining the variance of tablet CQAs over time. The Level 2 DR implementation was found to be necessary to reconcile noisy sensor measurements of CQA used in the process control systems. [5, 6] Furthermore, it was also demonstrated that, under predetermined nominal material variances, the Level 0 control commonly used in batch manufacturing showed promising control performance due to the reliable mechanical design of the equipment. Under conservative tuning, both Level 1 and Level 2 active process control schemes achieved comparable promising performance in the presence of inherent system and sensor noise. However, when materials were subject to significant variations in, e.g., bulk density, the active process control with Level 1 and 2 exhibited better performance. In addition, the Level 2 MPC control showed better performance than the Level 1 PID loops in process automation roles. For example, Level 2 MPC can rapidly reach a CQA set point during process start-up.

The results of this work conclusively demonstrate that an active process control strategy, with process knowledge [7] and advanced model-based control techniques, is indispensable to the QbD implementation of pharmaceutical continuous manufacturing, and that it ensures more robustness and efficiency than the techniques used in conventional batch manufacturing, providing the key ingredient towards the new paradigm of Quality-by-Control (QbC). [8] **References**

1. Ierapetritou M, Muzzio F, Reklaitis G. Perspectives on the continuous manufacturing of powder-based pharmaceutical processes. *AIChE Journal*. 2016;62(6):1846-1862.

2. Yu LX, Amidon G, Khan MA, et al. Understanding pharmaceutical quality by design. *The AAPS Journal*. 2014;16(4):771-783.

3. Diab S, Gerogiorgis DI. Process modelling, simulation and technoeconomic evaluation of crystallization antisolvents for the continuous pharmaceutical manufacturing of rufinamide. *Computers and Chemical Engineering*. 2018;111:102-114.

4. Su Q, Moreno M, Giridhar A, Reklaitis GV, Nagy ZK. A systematic framework for process control design and risk analysis in continuous pharmaceutical solid-dosage manufacturing. *Journal of Pharmaceutical Innovation*. 2017;12:327-346.

5. Moreno M, Liu J, Ganesh S, et al. Steady-state data reconciliation of a direct compression tableting line. Paper presented at: AIChE Annual Meeting, 2017; Minneapolis, US.

6. Liu J, Su Q, Moreno M, Laird C, Nagy Z, Reklaitis G. Robust state estimation of feeding-blending systems in continuous pharmaceutical manufacturing. *Chemical Engineering Research and Design*. 2018:Accepted.

7. Su Q, Bommireddy Y, Gonzalez M, Reklaitis GV, Nagy ZK. Variation and risk analysis in tablet press control for continuous manufacturing of solid dosage via direct compaction. Paper presented at: The 13th International Symposium on Process Systems Engineering PSE 2018, 2018; San Diego, CA.

8. Koswara A, Nagy ZK. On-Off feedback control of plug-flow crystallization: a case of Quality-by-Control in continuous manufacturing. *IEEE Life Sciences Letters*. 2017;3(1):1-4.

AUTHOR INDEX

AUTHOR INDEX

AUTHOR INDEX

AUTHOR INDEX

AUTHOR INDEX

AIChE
120 Wall Street, FL 23
New York, NY 10005-4020

CURRAN ASSOCIATES INC.
proceedings.com

ISBN 978-1-5108-7622-4
9 781510 876224